Sensors 2018

Topical Conference at the 2018 AIChE Annual Meeting

Pittsburgh, Pennsylvania, USA
28 October - 2 November 2018

ISBN: 978-1-5108-7635-4

Printed from e-media with permission by:

Curran Associates, Inc.
57 Morehouse Lane
Red Hook, NY 12571

Some format issues inherent in the e-media version may also appear in this print version.

Printed by Curran Associates, Inc. (2019)

For permission requests, please contact AIChE
at the address below.

AIChE
120 Wall Street, FL 23
New York, NY 10005-4020

Phone: (800) 242-4363
Fax: (203) 775-5177

www.aiche.org

Additional copies of this publication are available from:

Curran Associates, Inc.
57 Morehouse Lane
Red Hook, NY 12571 USA
Phone: 845-758-0400
Fax: 845-758-2633
Email: curran@proceedings.com
Web: www.proceedings.com

TABLE OF CONTENTS

(134a) An Optical Near Infrared Doxorubicin Sensor Discovered By Spectroscopic and Chemometric Analysis of Nanosensor Libraries ... 1
Jackson Travis Del Bonis-O'Donnell, Rebecca Pinals, Sanghwa Jeong, Ami Thakrar, Russ Wolfinger, Markita Landry

(134b) Toehold-Mediated DNA Strand Displacement Reactions for Quantitative Paper-Based Diagnostics ... 2
Elizabeth Phillips, Taylor Moehling, Jacqueline Linnes

(134c) Integration of Surface-Enhanced Raman Scattering and Dielectrophoresis for Rapid Separation and Detection of Bacteria in Real-Time ... 3
Qiuming Yu, Daniel David Galvan

(134d) Non-Invasive Plasmonic Biosensors for in Situ Glucose Monitoring ... 4
Nihan Yonet-Tanyeri, Ji Eun Park, Richard P. Van Duyne, Milan Mrksich

(134e) Effects of Low Dose Ionizing Radiation on Microorganisms for Creating Inconspicuous Biosentinels ... 5
Molly Wintenberg, Lisa Manglass, Nicole Martinez, Mark Blenner

(134f) Bio-Conjugated, Single-Use Biosensor for the Detection of Biomarkers of Prostate Cancer ... 6
Yifan Dai, Jiwei Yao, Yuan Wang, Chung-Chiun Liu

(134g) Sedimentation Separation of Red Blood Cells and Bacteria for Rapid Diagnosis of Blood Infections ... 7
William G. Pitt, Mahsa Alizadeh, Ryan L. Wood, Alex K. Hunter, Rebekah N. Torgesen

(134h) Integrated Point of Care Device for Nucleic Acid Extraction, Isothermal Amplification, and Fluorescence-Based on-Line Detection of Viral Genetic Material ... 8
Jackelin Mendoza-Ramos, Everardo Gonzalez-Gonzalez, Andres Gracia-Rubio, Grissel Trujillo-De Santiago, M. M. Alvarez, Sergio Omar Martinez-Chapa

(178a) Non-Invasive Disease Diagnosis Using Wearable Sensing Technologies ... 9
Hossam Haick

(178b) Molecular Technologies for Robust Detection of Proteins in Bodily Fluids ... 10
Hadley D. Sikes

(178c) Exploiting Oxygen Inhibited Photopolymerization to Control Shape, Size and Network Architecture of Functional Hydrogels As a Biosensing Platform ... 11
Katie Dongmei Li-Oakey

(178d) Microelectrode Cholesterol Sensing at Single Cells, Animal Tissues, and the Human Mucosa for Pre-Clinical Studies and Patient Evaluations ... 12
James Burgess

(231a) Invited: Nano-Structured Materials Enabled High-Temperature Gas Sensors: From Resistor-Type Sensors to Passive SAW Sensors ... 13
Yu Lei

(231b) Invited: Microprobe for Sensing of Multiple Neurochemicals In Vivo ... 14
Harold G. Monbouquette

(231c) Enzyme-Conjugated Nanosensors with a Tunable Detection Limits for Small Bio-Molecule Monitoring ... 15
Mark S. Ferris, Makayla K. Elms, Kevin J. Cash

(231e) Array of Nanostructured Electrode Tailored from Isolated to Continuum Monolayer for Chemical Sensing at Sub Parts per Trillion ... 16
Jennifer A. Arcila, Rahul Tevatia, Ravi Saraf

(231g) Developing High Performance Low Cost Ammonia Sensors Based on a Substrate-Directed Solution Crystallization Process ... 17
Xuecheng Yu, Mohamed Kilani, Evan Schaefer, Guangzhao Mao

(231h) Portable and Low-Cost Potentiostat System for Quantification of Cadmium in Wastewaters ... 18
Crhistian Camilo Segura, A. L. C. Perilla, S. L. F. Gonzalez, Mabel Juliana Noguera Contreras, Juan C Cruz, Johann F Osma

(265a) Integrated Biosensor for Rapid and Point-of-Care Sepsis Diagnosis ... 20
Jouha Min, Fillip Swirski, Hakho Lee, Ralph Weissleder

(265b) Development of a Fast-Responding, Minimal-Equipment Biosensor for Zinc Deficiency ... 21
Monica McNerney, Mark P. Styczynski

(265c) Non-Invasive Cell Density Measurement of Mammalian Cell Cultures in Early Stage Seed Trains 22
Jana McGuin, Sarah Magnino, Mark Berge, Michael Mollet
(265d) Protein Detection with Peptoid-Functionalized Carbon Nanotube Optical Sensors 23
Linda Chio, Jackson Travis Del Bonis-O'Donnell, Mark A. Kline, Ronald N. Zuckermann, Markita Landry
(265e) High-Throughput Single Cell Analysis of Deubiquitinating Enzyme Activity in Intact Cells 24
Manibarathi Vaithiyanathan, Nora Safa, Shayan Sombolestani, Adam Melvin
(265f) A Handheld Optical Detection Method for Detection of Vibrio Cholerae in Environmental Water Samples 25
Katherine N. Clayton, Julia G. Fraseur, Dong Hoon Lee, Taylor Moehling, Steven T. Wereley, Jacqueline C. Linnes, Tamara L. Kinzer-Ursem
(265g) Phenome to Genome Enabled By Microfluidics and High-Throughput Quantitative Microscopy 26
Hang Lu
(279a) Control of Pseudomonas Aeruginosa Biofilms By Electrical Currents Using a Simple Agar Model 27
Devendra Dusane, Varun Lochab, Travis Jones, Casey Peters, Amitava Das, Sashwati Roy, Chandan Sen, Vish Subramaniam, Daniel Wozniak, Shaurya Prakash, Paul Stoodley
(279b) Prevention of Select Eskape Pathogens from Attaching to Titanium Using Cathodic Voltage Controlled Electrical Stimulation Combined with Antibiotic Therapy 28
Mary Canty, Nicole Luke-Marshall, Anthony Campagnari, Mark Ehrensberger
(279c) Computational Modeling of Cathodic Voltage Controlled Electrochemical Treatment of Biofilms in-Vivo 29
Amir Mokhtare, Mark Ehrensberger, Edward P. Furlani
(279d) Electroactive Surfaces and Their Use for Biofilm Removal to Advance Wound Healing 31
Abdelrhman Mohamed, Hannah M. Zmuda, Mia Mae Kiamco, Ahmed Ben Sahil, Yash Raval, Douglas R. Call, Robin Patel, Haluk Beyenal
(279e) Toward to the Design of an Electrochemical Therapy (ECT) Against Microbial Infection 32
Nna-Emeka Onukwugha, Eloise Parry-Nweye, Tagbo H. R. Niepa
(279f) Wireless Electrostimulation to Eradicate Bacteria Biofilm 33
Hao Wang, Dacheng Ren
(279g) Electrochemical Detection of Bacterial Biofilms on Titanium 34
Caelen Clark, Mark Ehrensberger
(279h) Novel Focused Multivector Ultraviolet (FMUV) Disinfection without Manual Cleaning and Chemical Disinfection in-between Surgeries and throughout the Hospital Environment 36
Donna Armellino, Luis F. Romo, Thomas J. Walsh, Vidmantas Petraitis, Audrey McNicholas, Wladyslaw Kowalski, Mao-Wen Weng
(292a) Boronate Ester-Based Dynamic Nucleic Acids for Templated Analyte Detection 37
Heidi R. Culver, Kelly Kepler, Christopher N. Bowman
(292c) Mechanistic Optimization of Floating Gate Transistors for Biosensing Applications 38
Mathew Thomas, Kevin D. Dorfman, C. Daniel Frisbie
(292d) Computational Optimization of Metal-Organic Framework (MOF) Arrays for Chemical Sensing 39
Jenna Gustafson, Christopher E. Wilmer
(292e) Morpholino Materials for Diagnostic Applications 40
Sade Ruffin, Eshan Treasurer, Isabella Hung, Rastislav Levicky
(292f) Cardiac Troponin I Detection Using Antibody-Immobilized Disposable Cover Glass and AlGaN/GaN High Electron Mobility Transistors 41
Jiancheng Yang, Patrick Carey IV, Fan Ren, Yu-Lin Wang, Michael L. Good, Soohwang Jang, Michael A. Mastro, Stephen J Pearton
(292h) Nanostructured Polymeric Membranes for in-Situ Measurement of Exhaled Formaldehyde and Acetone Kinetics As Early-Stage Non-Invasive Markers of Lung Disease 42
Anastasios Angelopoulos, Ulzii Badmaarag
(321a) Invited Talk: Advancing Biosensing with Hybrid Nanomaterials and Machine Learning 43
Alexander Star
(321b) Viscosity Measurement at the Point-of-Need: A Smartphone Capillary-Based Approach 44
Jose C. Contreras-Naranjo, Vijetha Nagendra Prakash, Xiaorui Dong, Victor M. Ugaz
(321c) Corona Phase Molecular Recognition Sensors in Marine Organisms for Physiological Biologging: A Feasibility Study 45
Michael A. Lee, Nathan Chan, Freddy T. Nguyen, Naveed Bakh, Kelvin K. Jones, Crystal Pham, Pablo Garcia-Salinas, Daniel Garcia-Parraga, Vicente Marco, Michael Strano
(321d) Detection of Beta Carotene and Lutein Using Electrochemical Impedance Spectroscopy 46
Sabrina Marnoto, Jeffrey M. Halpern

(321e) A New Surface Functionalized Biosensor for Long-Term In Vivo Glucose Monitoring 47
Yikun Huang, Yi Luo, Haomin Liu, Donghui Song, Qiuchen Dong, Jing Zhao, Yu Lei
(321f) Mobile Technology Based ECL Biosensor Instrumentation 48
Daniel Marsh, Hyun J. Kwon
(321g) Electrochemical Detection of Extracellular Bacterial Compounds Using Capillary Electrophoresis 49
Martin K. Kimani, Edgar D. Goluch
(388a) Invited Talk: The Next Dimension of Detection: Biomechanical Analysis of Tissue 50
Andrea M. Armani, Alexa Hudnut, Lili Lash-Rosenberg, An Xin, Juan Doblado, Cecilia Zurita-Lopez, Qiming Wang
(388b) Rapid Biosensing of Endocrine Disruptors with Cell-Free Protein Synthesis 51
Bradley C. Bundy, J Porter Hunt, Seung Ook Yang, Miriam Shakalli Tang, David W. Wood
(388c) Lab-on-Skin: Epidermal Microfluidic Device for the Capture, Storage, and Colorimetric Sensing of Sweat 52
Yi Zhang, John A. Rogers
(388d) Stimulus Response Characterization of an Elastin-like-Polymer Modified Surface for Biosensor Applications 53
Marissa Morales, Eva Rose M. Balog, Jeffrey M. Halpern
(388e) Development of a Unique Dual Ionophore Ion Selective Electrode for the Detection of Proteins and Cells 54
Olivia Reynolds, Xuesong Li, Bernard J. Van Wie
(388f) Evaluation of Aptamer Technology for Detection of Quorum Sensing Molecules Produced By Pseudomonas Aeruginosa 55
Pranali Buch, Edgar D. Goluch
(388g) Chromogenic Ethanol Sensors Enabled By Multi-Stimuli-Responsive Shape Memory Polymers 56
Abdullateef Gari, Peng Jiang
Author Index

Some abstracts in this book refer to images that were not made available by the publisher

536612 An Optical Near Infrared Doxorubicin Sensor Discovered By Spectroscopic and Chemometric Analysis of Nanosensor Libraries

Abstract:

Polymer-functionalized single walled carbon nanotubes (SWNTs) are a promising new class of near-infrared (NIR) fluorescent nanomaterials with applications to non-invasive *in vivo* biomolecular sensing. The adsorbed polymer-phase can impart specific molecular recognition, whereby analyte binding can be monitored by measuring the resulting modulation of SWNT fluorescence. Typical preparations of polymer-functionalized SWNT nanosensors are composed of a mixture of up to fifteen nanotube chiralities, each with a unique fluorescence emission and potentially unique response to an analyte. Currently, methods for screening new nanosensor-target pairs involve searching for broad changes in fluorescence intensity and do not capture the unique fingerprint of the combined nanosensor chiralities. Herein, we demonstrate rapid and quantitative analysis of fluorescence emission spectra collected from chirality-mixed SWNT-polymer conjugate library screens with distance metric calculations, hierarchical clustering (HC) and principal component analysis (PCA). SWNT-polymer conjugates each functionalized with a different polymer corona phase, including polynucleotides, phospholipids, and amphiphilic heteropolymers, are screened against a panel of >40 biological targets. Our analysis quickly identifies nanosensor responses to difference analytes, including changes in intensity and shifts in peak wavelength, using multivariate approaches without the need for time-intensive non-linear regression fitting of individual spectral peaks. Analytes that induce similar responses, e.g. catecholamines, are grouped using hierarchical clustering, demonstrating that our approach can enable classification of nanosensor-target interactions. Our approach further identifies a novel NIR nanosensor for the detection of the chemotherapeutic doxorubicin (DOX) with μM sensitivity. We find that the kinetics of binding as well as the optical response of nanosensors to DOX is strongly dependent on the polynucleotide sequence used to functionalize the SWNT. Furthermore, we show that the fluorescence intensity increases and blue-shifted emission of the nanosensor is a combination of an interaction between the DOX and the SWNT surface as well as a perturbation to the conformation of DNA in its corona-phase. Finally, we demonstrate imaging of DOX distribution in biological tissues, and extrapolate its use to image drug bio-distributions *in vivo*.

533560 Toehold-Mediated DNA Strand Displacement Reactions for Quantitative Paper-Based Diagnostics

Abstract:

Molecular techniques that specifically and quantitatively detect low copy numbers of nucleic acids are essential for identifying pathogens in patient and environmental samples and informing their treatment. Isothermal loop-mediated amplification (LAMP) is a promising nucleic acid detection method because it rapidly and exponentially amplifies dilute template DNA, performs robustly with complex samples, and requires minimal equipment. Primers are commonly tagged for post-amplification detection on a simple, lateral flow sandwich-based immunoassay (LFIA) that yields colored bands [1]. However, without an electronic reader, LFIAs provide only a qualitative readout of nucleic acid concentration. Moreover, we have found that the color intensity of the test bands significantly affects users' accurate interpretation of even qualitative LFIA results [2]. Therefore, here we explore molecular mechanisms (toehold-mediated strand displacement) that could 1) permit more accurate qualitative assessment and 2) provide a quantitative readout of nucleic acids on paper-based test strips.

We recognized that tagging LAMP primers for subsequent LFIA detection frequently yields false positive results. Primer-dimers can cause spurious amplification products that are captured on the LFIA, causing widely varying test band intensities that are difficult for users to accurately interpret. To improve users' instrument-free interpretation of paper-based diagnostics, we first incorporated toehold-mediated, strand displacement probes into LAMP [3]. We demonstrate that strand displacement probes prevent off-target LFIA capture caused by primer dimerization, thereby eliminating false positive results and permitting definitive interpretation of traditional, LFIA results. We find this simple modification to LAMP provides robust LFIA readout of multiple bacterial targets (*Vibrio cholerae* and *Escherichia coli*), even in the presence of complex sample matrices (pond water and human serum).

Second, we explored the utility of toehold-mediated strand displacement reactions to supplant the post-amplification LFIA for quantitative detection of pathogens' DNA on paper strips. We printed low-cost DNA-hybrid hydrogels, functionalized by the same strand displacement probes described above, into paper strips. In the absence of a target pathogen's DNA, we found that the hydrogel prevents flow through the paper strip; in the presence of the pathogen's DNA, the hydrogel releases fluid. This unequivocally observed flow/no-flow response to pathogenic DNA could simplify diagnosis of infectious diseases on paper test strips. Furthermore, we demonstrate the hydrogel functionalized paper strips' potential to yield a quantitative read-out by correlating DNA concentration with flow distance. In contrast to LFIA's qualitative results, hydrogel functionalized paper strips offer an instrument-free, quantitative detection of pathogenic DNA. The integration of toehold-mediated DNA strand displacement reactions with paper strips constitutes a significant step towards clinically impactful pathogen detection outside of laboratories.

[1] Rodriguez, et al. *Anal. Chem.*, 2015, 87 (15), pp 7872–7879.

[2] Phillips, et al. *Anal. Chem.* (in review).

[3] Yu, et al. *Anal. Chem.*, 2015 (87), pp 3314-3320.

532278 Integration of Surface-Enhanced Raman Scattering and Dielectrophoresis for Rapid Separation and Detection of Bacteria in Real-Time

Abstract:

The need for rapid and specific pathogen detection is of utmost importance. Current culture based methods are time consuming, and more efficient technologies would benefit wide-ranging fields including food safety, biomedicine, homeland security, and environmental monitoring. Surface-enhanced Raman scattering (SERS) can afford real-time interrogation of the unique biochemical composition of pathogenic bacteria. Species-level differentiation is enabled by the differences present in the bacterial cell wall. However, SERS measurements are typically conducted with concentrated samples (10^8 CFU/mL) of pure cultures that are far outside the clinically relevant range. Real-world samples are often dilute (< 10^3 CFU/mL), and present in bacterial mixtures and/or complex media. The objective of this work was to develop SERS-based approach to enable real-time detection of dilute and/or mixed bacterial samples. To combat these issues, we have developed a two-pronged SERS approach using: (1) long range SERS (LR-SERS) devices to acquire SERS characteristics of the subsurface cellular makeup and (2) SERS devices integrated with dielectrophoresis (DEP) for the concentration and separation of bacterial cells.

We conducted finite-difference time-domain (FDTD) electromagnetic simulations to design SERS-active substrates with extended strong electric fields (i.e., hot spots) to dielectric medium. To achieve this, SERS-active Au nanohole arrays (NHAs) were embedded in refractive index-matched environments. FDTD simulations were also conducted to investigate the plasmonic response of NHAs in symmetric and asymmetric dielectric environments. The optimal structures were determined for both circular and x-shaped NHAs and fabricated via nanofabrication methods. The SERS signals of surface bound analytes dramatically increased when placed in the refractive index-matched environment. Furthermore, SERS signals were observed at a distance of 10 nm from the nanohole array surface. The increased penetration depth could enable examination of the unique bacterial composition between the peripheral cell wall and cytoplasm.

We incorporated DEP into the SERS biosensor by dual-function, nanostructured electrodes. "Point-and-plate" and "interdigitated" electrode configurations were studied with respect to SERS detection performance. Generally, bacteria localize on the sensing surface in regions with high or low electric field gradients depending on the unique cellular dielectric properties. The effect of the applied AC frequency on the capture efficiency and SERS signals of bacteria was investigated for the point-and-plate configuration. Application of DEP afforded the successful detection of 10^5 CFU/mL *E. coli* solutions, and the applied electric fields did not alter the SERS spectra of Gram-positive and Gram-negative bacteria. Interdigitated electrodes contain periodic regions with high and low electric field gradients at the edge and center of the electrode, respectively, and can therefore be used to separate bacterial mixtures at either the edge or center of the electrodes. The concentrated and separated bacterial mixtures can be directly detected using SERS.

The SERS devices developed here, used in conjunction with portable Raman spectrometers, could offer

frontline diagnostic solutions to simultaneously separate, concentrate, and differentiate dilute bacterial mixtures in real-time.

533957 Non-Invasive Plasmonic Biosensors for in Situ Glucose Monitoring

Abstract:

Careful management of blood glucose level via frequent monitoring for diabetic patients is very important especially in the absence of a cure. The electrochemical detection based technologies remain the main focus of widely used commercially available finger stick testings for monitoring blood glucose level. However, the technological advancements have opened new avenues for alternative methods that will bring the ease of use and comfort for frequent users. Especially, label-free optical approaches have attracted a considerable attention. Here with this work, we propose a non-invasive plasmonic biosensor design for in situ glucose monitoring from the body fluid (interstitial fluid) using Surface Enhanced Raman Spectroscopy (SERS). The plasmonic device is composed of an array of polymeric microneedles and integrated with plasmonic component, metal nanoparticles, that are localized on the microneedles surface. The plasmonic component is chemically modified with self-assembled monolayer of glucose capture ligand. In situ sensing mechanism relies on capturing glucose from interstitial fluid on the metal nanoparticle surface once the device is pierced in the skin and real-time detecting SERS fingerprints of the capture ligand for quantitative analysis. Our initial findings showed that the plasmonic biosensor is SERS active when the device is inserted in a synthetic skin mimicking model (skin phantom). The highest SERS activity was observed at the microneedles tip. It is very important to show the mechanical and functional stability of the plasmonic device upon piercing to the skin and under exposure to the body fluid. The stability tests demonstrated that the proposed plasmonic device maintains its structural integrity and confirmed mechanical stability upon multiple piercing in the skin phantom. Additionally, metal nanoparticles stay intact on the polymer microneedles as the device is kept in the body fluid mimicking buffer condition over a month. Finally, the biosensing capability of the developed plasmonic device as well as its continuous glucose monitoring capacity will be discussed.

534957 Effects of Low Dose Ionizing Radiation on Microorganisms for Creating Inconspicuous Biosentinels

Abstract:

Traditional radiation detection systems are extremely sensitive to low doses of ionizing radiation, however the disadvantages of being easily identifiable, mandatory placement near a radioactive source for detection, and capacity to report radioactivity at a specific moment in time diminish their applications in the field. Given the current state of world affairs, there is a need for an inconspicuous sensor capable of unattended autonomous operation for reporting radioactivity even if the source has been relocated. A set of transcriptional based biosensors capable of discerning radionuclide type have the potential to monitor and report on nuclear fuel cycle, enrichment, and weapon development activities in diverse environmental conditions. Transcriptional responses of model and environmental microorganisms exposed to acute and chronic alpha, beta, gamma and neutron radiations are analyzed through RNA sequencing to identify unique radiation-induced signatures. After quantitative PCR verification, these unique signatures could be used to engineer a set of microbial transcriptional based biosensors. Our work with *Pseudomonas putida* and *Escherichia coli* indicate key differences in gene expression after acute and chronic exposure to Plutonium 239 at a dose rate of 8.6 mGy/d. Substantial differences in differentially expressed genes were noted between acute and chronic exposures of *P. putida* indicating the potential presence of a radiation-induced time signature. Unique upregulated genes from each organism and time point will be engineered into initial toehold switch biosensor prototypes. Furthermore, changes in oxidative stress response were measured even due to low dose radiation exposure, suggesting one possible mechanism for radiation response. The results of this work will expand upon the limited knowledge of low level radiation effects in microorganisms and lead to passive and autonomous monitoring of clandestine nuclear activities.

528752 Bio-Conjugated, Single-UseÂ Biosensor for the Detection of Biomarkers of Prostate Cancer

Abstract:

Prostate cancer is prevalent among the cancers of men. Detection and screening of a reliable biomarker of prostate cancer is important for individual health of men. Prostate specific antigen (PSA) has been a commonly recognized as biomarker for prostate cancer, in spite of its false-positive limitation. On the other hand, alpha-methylacyl-CoA racemase (AMACR), a metabolic enzyme, which has been proven to be a highly expressed biomarker in prostate cancer cells. Therefore, a method or tool can detect either PSA or AMACR or both simply, cost effectively and with high sensitivity and selectivity is desirable. We describe a novel bio-conjugated, single use biosensor capable of detecting both PSA and AMACR antigens in undiluted human serum. The preparation of the biosensor by bio-conjugation mechanism was a rapid process, which could be completed within one day prior to actual testing. The bio-conjugation mechanism and the coverage of the electrode surface of the biosensor were experimentally assessed. Measurements of PSA and AMACR antigens and the specificity of the biosensor were carried out using differential pulse voltammetry (DPV). This study demonstrates a rapid, accurate, single-use and cost effective biosensor for detection of prostate cancer.

535952 Sedimentation Separation of Red Blood Cells and Bacteria for Rapid Diagnosis of Blood Infections

Abstract:

Fast and efficient separation of bacteria from blood is the first step towards the rapid diagnosis of blood infections. We have developed a spinning hollow disk that separates blood cells from bacteria by capitalizing on the differences in sedimentation velocities of these particles. Hollow disks of various size and design were constructed to examine hypotheses relating to optimal collection of bacteria from blood and favorable separation of blood cells from the plasma. Disk design characteristics included the disk diameter, shape of the weir and trough, baffles, and fillets in corners. Blood processing characteristics included the spinning speed and duration, and the deceleration of the disk. Measured parameters were the volume of separated plasma, and the concentration of bacteria and red blood cells in the separated plasma. The data were correlated by a parameter of characteristic sedimentation length, which is the length that an idealized red blood cell would travel during the entire spin. The data from various disk sizes and spinning speed profiles were nicely correlated with this parameter. The results show that a characteristic sedimentation length of 20 to 25 mm produces an optimal collection of bacteria in plasma. This corresponds to spinning a 12-cm-diameter disk at 3,000 rpm for 13 seconds. Following the spin, a careful and slow deceleration preserves the separation of cells from plasma, and provides a bacterial recovery of around 60%. This is the first step in a diagnostic tool designed to identify bacteria in blood infections. Subsequent steps include lysing and recovery of bacterial DNA, hybridization of sequences identifying species and antibiotic resistance genes, fluorescent labeling, and optofluidic detection of the labeled DNA.

536800 Integrated Point of Care Device for Nucleic Acid Extraction, Isothermal Amplification, and Fluorescence-Based on-Line Detection of Viral Genetic Material

Abstract:

Early diagnosis of viral diseases is vital for accurate treatment and containment of infections, and it leads to better outcomes for the patient and effective protection for the community. Motivated by recent epidemics (i.e., 2014 Ebola West Africa and 2015-16 Zika outbreaks), the World Health Organization has called for rapid, safe, and cost-effective virus diagnostic systems to be used at point-of-care (POC) locations.

Nucleic acid amplification methods have the accuracy, sensibility, and specificity needed to make them the tools of choice for viral diagnostics at early infection stages. However, conventional amplification techniques are poorly suited for POC application. Molecular diagnostic usually consists of three steps: genetic material extraction, amplification, and detection. Typically, the accomplishment of these three steps demands well-trained technical staff and costly reagents and equipment (i.e., an extraction reagent kit, a thermocycler, an electrophoresis chamber, and a trans-illuminator). These requirements increase diagnostic expenses and turnaround time, thereby limiting testing availability, especially in underprivileged or remote areas.

Herein, we present an integrated, three-stage, one-pot strategy to solve these drawbacks. In the first stage, a single and short high-voltage electric pulse releases viral nucleic acid material confined within infected cells, making it accessible for amplification. In the second stage, the genetic material is amplified in the same container by an isothermal amplification method, namely Recombinase Polymerase Amplification (RPA). Finally, the amplification product is detected by addition of GelRed® DNA-dye to the reaction mix, excitation with 360 nm UV light, and photodiode detection of the emitted fluorescence. All these steps are performed in a simple, cost effective, and portable device. Moreover, the on-line quantitative detection of the amplification product can be conducted using a simple Arduino-based module. We have demonstrated the use of this integrated device by quantitative detection of synthetic nucleic acid from Ebola, Zika, and influenza viruses released from bacterial and mammalian cells.

The minimal reagent usage and the lack of any requirement for specialized training, along with the simplicity, portability, and low cost of this device, make it suitable for cost-effective use at POC locations.

521368 Non-Invasive Disease Diagnosis Using Wearable Sensing Technologies

Abstract:

The lecture will present innovative technologies developed by our research group at the Technion. These technologies are non-invasive disease diagnosis tools that do not require needles or use of dangerous materials. The lecture will present, amongst other things, an electronic system that simulates the canine olfactory system and that detect diseases from exhaled breath. This system is known as â€œNano-Artificial Noseâ€. Clinical trials worldwide have shown that this system can detect 17 different diseases, including various cancers, neurological disorders and infectious diseases - all though a simple breath test. The lecture will also present the development and miniaturization of intelligent wearable technology that integrates functions similar to those in the human skin. This technology, known as the "Electronic Skin", has been found effective in monitoring a wide range of health indicators and help in disease diagnosis in both the developing and developed world. The link of these technologies to a unique information cloud, where the analysis of data takes place and results are sent back to the user or their personal physician, will be presented and discussed as well.

538565 Molecular Technologies for Robust Detection of Proteins in Bodily Fluids

Abstract:

Infectious diseases worldwide are either undiagnosed or incorrectly diagnosed at a staggering rate because diagnostic techniques that are feasible in the US are not feasible worldwide. An analysis of this problem will be presented along with two molecular technologies that my lab has developed for the detection of proteins in bodily fluids. Engineered binding proteins for sample capture and detection and photo-redox catalyzed polymerization reactions for colorimetric signal amplification will be demonstrated to simultaneously meet performance metrics while also meeting practical constraints set forth by the World Health Organization.

538891 Exploiting Oxygen Inhibited Photopolymerization to Control Shape, Size and Network Architecture of Functional Hydrogels As a Biosensing Platform

Abstract:

Hydrogel microparticle and nanoparticle-based hydrogel biosensing platforms require delicate control over network architecture and the conjugation of biofunctional groups, which impacts their mechanical properties and molecular permeability and diffusivity. For most hydrogel particle fabrication methods that rely on photoinitiated chain polymerization, the impact of exposure conditions and photopolymerization kinetics on the resulting particle properties is not well described. In this talk I will introduce oxygen inhibited photopolymerization within emulsion droplets as a high throughput technique for the controlled production of hydrogel particles. Empirical data, coupled with a detailed reaction-diffusion analysis, reveals that this technique can be used to exquisitely manipulate particle size, shape, and hydrogel particle properties. Different assays were conducted to examine the network properties of these particles, demonstrating a high degree of structural tunability, which, in turn, governs compound encapsulation and diffusion profile, as well as the presence of radial crosslinking gradients that regulate the availability of functional groups (Figure 1). A detailed quantitative description of the photopolymerization process using a reaction-diffusion model presents the ability to accurately predict and optimize hydrogel network properties to produce tailored hydrogel microparticles for targeted applications.

Figure 1: Oxygen-inhibited photopolymerization of droplets containing acrylate-PEG-biotin produces functional hydrogels with surfactant free surfaces. A. Schematic illustration of the fabrication of hydrogel microparticles. B. Pictograph of the process described in A, with Na-Rh fluorescence imaging showing the presence of crosslinking gradient in the radial direction.

531096 Microelectrode Cholesterol Sensing at Single Cells, Animal Tissues, and the Human Mucosa for Pre-Clinical Studies and Patient Evaluations

Abstract:

This research group has been investigating cholesterol diffusion rate using oxidase-modified microelectrodes positioned in contact with the cell plasma membrane (PM). The microelectrode-membrane interface establishes a thin aqueous layer between the two hydrophilic surfaces. Although only sparingly soluble, aqueous phase cholesterol between the microelectrode surface and the PM surface is enzymatically consumed perturbing the equilibrium between membrane cholesterol and aqueous cholesterol thus causing unidirectional efflux of cholesterol from the membrane only at the site of microelectrode contact. Lateral radial diffusion of cholesterol in the PM to the electrode contact site provides a steady-state efflux rate and microelectrode current response. *In vitro* analysis at cells and mouse tissues is conducted by repositioning the microelectrode between the contact position and a position away from the membrane surface so that the steady-state current response for contact can be referenced to the baseline current collected when the microelectrode is not contacting the membrane. For *in vivo* analysis at the human mucosa, a double potential pulse method was developed for data acquisition. Enzymatically generated hydrogen peroxide is allowed to accumulate between the microelectrode and the membrane during a 5 s hold time after which the hydrogen peroxide is consumed by oxidation during a 0.25 s potential pulse. A second replicate pulse is applied 0.5 s after the first pulse to gauge the background. The difference charge between pulse one and pulse two is collected serially and is averaged to produce the final signal. Key attributes of the method are that baseline drift of signal is avoided through the continuous background subtraction, differences in microelectrode properties are normalized by taking of the difference charge, and only hydrogen peroxide that accumulates at the electrode surface is observed in the difference charge measurement. Because the time at rest potential between pulses (quiet time: 0.5 s) is held longer that the pulse time (0.25 s), the difference charge is blind species that react at the electrode during the pulses through a diffusional flux, including any bulk hydrogen peroxide. These microelectrode cholesterol measurements have been used to demonstrate increased PM cholesterol diffusion rates at cell line models of cystic fibrosis (CF) and at trachea and nasal tissues for mouse models of the CF disease state relative to WT controls. More recently, the double potential pulse-microelectrode method was implemented to show that CF human bronchial epithelial cells exhibit increased PM cholesterol diffusion and that cholesterol diffusion at the oral mucosa of CF subjects is higher compared to healthy control subjects. These data demonstrate the clinical utility of this approach for both pre-clinical studies and for *in vivo* assessment of CF.

536340 Invited: Nano-Structured Materials Enabled High-Temperature Gas Sensors: From Resistor-Type Sensors to Passive SAW Sensors

Abstract:

In-situ, real-time monitoring the composition of combustion gases usually requires sensors to be operative at high temperature environment. Up to date, commercially available sensor technology for such harsh environment is extremely limited due to the high requirements for sensing materials and sensor performance. Therefore, there is an urgent need to advance high temperature stable sensing materials and develop novel high temperature sensing strategy to realize fast, sensitive, selective, rugged, and cost-effective high-temperature gas sensors for power and fuel systems. This talk will first discuss nanomaterials enabled high-performance resistor-type gas sensors, which have been extensively studied for high temperature environment in our group, followed by the discussion of its current challenges. Finally, we will discuss our recent work on exploring the strategy such as passive SAW sensor to improve the high temperature gas sensing wireless capability.

535143 Invited: Microprobe for Sensing of Multiple Neurochemicals *In Vivo*

Abstract:

Efforts to understand how neurons within networks interact to control behavior will be greatly facilitated by means with which to measure multiple neuroactive molecules in the brain simultaneously and in near-real time. We are developing an implantable microprobe capable of simultaneous rapid monitoring of three neurotransmitters/neuromodulators: dopamine (DA), glutamate (Glut), and acetylcholine. A detailed mathematical model is guiding electroenzymatic sensor optimization, and the use of a nanostructured Pt deposit has led to improved microsensors for choline and DA operated in constant potential amperometry mode. The creation of microsensor arrays for combined sensing of multiple neurochemicals is facilitated by the use of microcontact printing to deposit specific enzymes on selected microelectrode sites with fine control. Evaluation of probe performance *in vivo* has focused on the striatum and the basolateral amygdala (BLA). Optically evoked Glut release has been observed successfully in the BLA of rats, but not in those genetically modified animals whose pre-synaptic activity is subjected to inhibition. Also, the improved DA sensors have enabled detection of electrical stimulation-evoked DA response in the rat dorsal striatum in the drug-naïve state. Work is ongoing to correlate neurotransmitter release events with rat behavioral actions, which will provide valuable information pertinent to the search for therapeutic interventions for multiple neurological and psychiatric disorders.

531812 Enzyme-Conjugated Nanosensors with a Tunable Detection Limits for Small Bio-Molecule Monitoring

Abstract:

Analyte monitoring in biological settings is crucial for medical providers and researchers in numerous fields. However, the current methods of analyte detection can be limited in biocompatibility, their ability to measure continuously, monitor spatially, or provide non-invasive online monitoring *in vivo*. Nanosensors are a next generation sensor technology that hold the possibility of analyte monitoring with none of these limitations. Their miniaturized nature allows them to provide analyte information through a three dimensional space while their light-based information transfer makes them ideal for continuous monitoring. Therefore, nanosensor technology development can lead to more accurate medical diagnoses and disease management and allow for cell culture analysis in ways not currently possible. However, functionalization of nanosensors is necessary to expand the range of accessible analytes. In this work, we develop a platform of enzyme-linked nanosensors (ELiNS) for glucose detection. The platform consists of a biotin/avidin based conjugation approach to attach the enzyme glucose oxidase to oxygen-detecting nanosensors. We show that the dynamic range can be shifted by adjusting amount of enzyme associated with each nanosensor, providing an easy tool to produce an optimized sensor for multiple applications. We also show that the enzyme/nanosensor ratio can be easily adjusted by controlling the amount of biotin-functionalized lipids on the nanosensor surface. Furthermore, we show that the specificity of the sensor can be easily adjusted by swapping the enzyme for a different type of oxidase enzyme, making the technique ideal for building a suite specific towards other small biomolecules such as lactic acid.

530663 Array of Nanostructured Electrode Tailored from Isolated to Continuum Monolayer for Chemical Sensing at Sub Parts per Trillion

Abstract:

Electrochemical sensors are the most ubiquitous chemical sensors. For example, glucometer, with worldwide market of over $2 billion is arguably the most successful biosensor in history. It is becoming evident that nanostructured sensing electrodes can enhance the performance of a device significantly due to their unique characteristic high mass transport rate compared to planar electrode. Fabrication of gold nanostructured arrays to enhance electrochemical reactions will be discussed. Metal scaffold on DNA or metal nanoparticles were used as nanostructures. The Electrochemical reactions are measured by differential reflectometry, a novel optical technique that probes fluctuations of the refractive index at the electrode/electrolyte interface. The electrochemical reactions were measured on a 6 μm diameter spot defined by the He/Ne laser beam. The refractive index at the electrode interface varies due to redox reactions. By modifying the surface from isolated nanoelectrode array to continuum nanostructured electrode, it was possible to control the nature of the diffusion layer. The catalytic action of the nanostructure and the nanoelectrode effect of diminishing diffusion were quantitatively explored as function of the modulation of the inter-nanostructure distance. Measurements of lead, arsenic and mercury in solution from parts-per-million to parts-per-trillion levels will be discussed.

531234 Developing High Performance Low Cost Ammonia Sensors Based on a Substrate-Directed Solution Crystallization Process

Abstract:

The major barriers in commercializing nano/micro-wire technologies are the high price and complicated manufacture methods. Due to these difficulties, very few nano/micro-wire sensors have reached the market in the last decade. This talk focuses on a patented substrate-mediated crystallization process to deposit conductive tetrathiafulvalene (TTF) and tetracyanoquinodimethane (TCNQ) nano/micro-wires directly on sensor substrates. TTF-TCNQ is often considered the first synthesized organic metal with metallic conductivity. This class of charge-transfer salts has the additional advantage of allowing fine tuning of chemical structure and conductivity by pairing an electron donor with different electron receptors or vice versa. Using electrochemical crystallization, we show that it is possible to deposit TTF or TCNQ nano/micro-wires directly across the photolithographic metal contact lines as the sensing element for impedance-based gas sensing. We are applying the charge-transfer salt assembly for ammonia sensing. Exposure to ammonia can cause irritation and serious burns on the skin and in the mouth, throat, lungs, eyes and even death at high levels. There is a market demand for high performance low cost ammonia sensors for wide area and real time monitoring. Our nanowire sensors are capable of detecting ammonia at sub ppm levels within a few minutes. We have achieved higher selectivity especially against humidity variation by using different nanowire materials as sub-units in a sensor array, collecting large amount of impedance spectral data in different gas mixtures, and applying machine learning algorithms to differentiate ammonia and water vapors. We are working with several companies to commercialize this novel gas sensing technology. This work is supported by the National Science Foundation, US Army Research Office, and MTRAC.

538389 Portable and Low-Cost Potentiostat System for Quantification of Cadmium in Wastewaters

Abstract:

The incessant growth of human activities has been accompanied by a constantly increasing release of pollutants to the environment, some of which tend to accumulate in air, water, and land to levels that pose a risk several living organisms and fragile ecosystems. One of such compounds is Cadmium, which is a relatively scarce metal usually found in the earth's crust and is also the inevitable byproduct of the extraction processes of zinc, lead and copper [1]. Even at very low concentrations, Cadmium is cytotoxic and carcinogenic to humans and due to its marked tendency to accumulate in kidneys, bones and the respiratory system [2][3]. Cadmium dissolves easily in polluted waters with low pH levels and can be transported to fresh superficial waters for human consumption. Furthermore, it could reach soil or fertilizers from which plants might absorb it to finally end up in food sources [4].

Cadmium in water sources is usually quantified via atomic absorption spectroscopy [5] and inductively coupled plasma atomic emission spectroscopy [6][7]. These two analytical methods are approved by federal agencies and organizations such as the US-EPA [8]. Additional analytical approaches based on electrochemical measurements are quite precise and include anodic stripping voltammetry [9], adsorptive cathodic stripping voltammetry [10], and potentiometric stripping analysis [8]. Main drawbacks of these methods include the pre-treatment of samples prior to measurements via tedious and lengthy protocols, and the need for expensive and large devices that limit in-situ and real-time monitoring [11]. Accordingly, there is a need for high sensitivity methods to detect and quantify cadmium in a simple, rapid, and inexpensive manner. An interesting avenue to respond to this challenge is to manufacture a portable device that also incorporates the sensitivity of electrochemical reactions. Such a device is known as a potentiostat and mainly consist of a stepped triangular signal generator, a transimpedance amplifier and a voltage follower. Here we benchmark our recently developed potentiostat with a commercially available potentiostat. Our manufacturing strategy allowed us to prepare a low-cost potentiostat. Details of the protocols can be found elsewhere [12]. Briefly, the portable device is assembled in an electronic PCB which have a composite material known as FR4 as substrate, and two coper layers on each side, it has dimensions of 5,34 cm x 4,45 cm and is capable of generating stepped voltage signals to conduct electrochemical measurements. Also, the device accuracy is of 0.8mV in the generated signal and is able to capture currents in the range of nA. The development was enabled by an ultra-low-power microcontroller from Texas Instruments (USA).

Cells of three electrodes were fabricated for measuring cadmium in water with both devices via cyclic voltammetry (CV). Measurements were successfully conducted with an exceedingly low sample volume of less than 300 µL. Materials such as copper (Cu) and Gold (Au) served as primary components for the developed cells and in some cases 4-Aminothiophenol (4-ATP) was covalently conjugated to the metal surface to increase the Cadmium chelation efficiency. Previous work with a commercial potentiostat demonstrated that sensors based on Cu, Au/Cu and Au/Cu modified with 4-ATP were able to measure Cadmium in aqueous solutions ranging from 13 Cd mg/mL to 0,786 Cd mg/mL. To carry out the tests, one cycle of CV between –1 V to 0 V was applied, which resulted in an anodic peak of cadmium oxidation between –0.5V and –0.4V. Current at this peak was directly proportional to the amount of substrate present in the sample. Different cell architectures will be tested to establish the detection limit of the device, which is expected to be in close proximity to the maximum allowable cadmium concentration in water for human consumption. Accordingly, our sensing technology appears suitable to monitor wastewater and water sources in a rapid, inexpensive, and portable manner. Moreover, the device is user-friendly and requires a low level of training for operation. We expect to develop avenues to use the device with samples of challenging physicochemical properties such as food, urine, saliva and blood.

REFERENCES

1] Lenntech. Cadmium (Cd), Chemical properties of cadmium - Health effects of cadmium - Environmental effects of cadmium. https://www.lenntech.com/periodic/elements/cd.htm#ixzz5CYn8XMpv . [Accessed: 14-Dec-2017]. 2017

[2] WHO. Exposure to cadmium: A major public health concern. http://www.who.int/ipcs/features/cadmium.pdf. [Accessed: 12-Apr-2018]. 2010.

[3] J. Godt, F. Scheidig, C. Grosse-Siestrup, V. Esche, P. Brandenburg, A. Reich, and D. A. Groneberg, "The toxicity of cadmium and resulting hazards for human health.," J. Occup. Med. Toxicol., vol. 1, p. 22, Sep. 2006. DOI: 10.1186/1745-6673-1-22

[4] WHO. Cadmium in Drinking-water Background document for development of WHO Guidelines for Drinking-water Quality. http://www.who.int/water_sanitation_health/dwq/chemicals/cadmium.pdf [Accessed: 12-Apr-2018]. 2011.

[5] R. Golbedaghi, S. Jafari, M. R. Yaftian, R. Azadbakht, S. Salehzadeh, and B. Jaleh, "Determination of cadmium(II) ion by atomic absorption spectrometry after cloud point extraction," J. Iran. Chem. Soc., vol. 9, no. 3, pp. 251–256, Jun. 2012. https://doi.org/10.1007/s13738-011-0018-7

[6] J. M. E. Almendro, C. B. Ojeda, A. G. De Torres, and J. M. C. Pavon, "Determination of Cadmium in Biological Samples by Inductively Coupled Plasma Atomic Emission Spectrometry After Extraction With," Elements, vol. 117, no. November, pp. 1749–1751, 1992.

[7] T. Daşbaşı, Ş. Saçmacı, A. Ülgen, and Ş. Kartal, "A solid phase extraction procedure for the determination of Cd(II) and Pb(II) ions in food and water samples by flame atomic absorption spectrometry," Food Chem., vol. 174, pp. 591–596, May 2015. https://doi.org/10.1016/j.foodchem.2014.11.049

[8] Agency for Toxic Substances and Disease Registry (ATSDR). Toxicological profile for Cadmium. Atlanta, GA: U.S. Department of Health and Human Services, Public Health Service. 2012.

[9] Rajawat, D.S., Kumar, N. & Satsangee, S.P. J Anal Sci Technol (2014) 5: 19. https://doi.org/10.1186/s40543-014-0019-0

[10] Deswati D, Pardi H, Suyani H, Zein R. Adsorptive Cathodic Stripping Voltammetric Methodwith Alizarin for the Simultaneous Determination of Cadmium, and Zinc in Water Samples. Orient J Chem 2016;32(6). Available from: http://www.orientjchem.org/?p=25331

[11] Zohreh Dahaghin, Paul A. Kilmartin, Hassan Zavvar Mousavi, Simultaneous determination of lead(II) and cadmium(II) at a glassy carbon electrode modified with GO@Fe3O4@benzothiazole-2-carboxaldehyde using square wave anodic stripping voltammetry, Journal of Molecular Liquids, Volume 249, 2018, Pages 1125-1132

[12] Segura CC, Osma JF (2017) Miniaturization of Cyclic Voltammetry Electronic Systems for Remote Biosensing. Int J Biosen Bioelectron 3(3): 00068. DOI: 10.15406/ijbsbe.2017.03.00068

524997 Integrated Biosensor for Rapid and Point-of-Care Sepsis Diagnosis

Abstract:

Purpose: Sepsis is an often fatal condition that arises when the immune response to an infection causes widespread systemic organ injury. A critical unmet need in combating sepsis is the lack of accurate early biomarkers that produce actionable results in busy clinical settings. Here, we report the development of a point-of-care platform for rapid sepsis detection. Termed IBS (integrated biosensor for sepsis), our approach leverages i) the newly-found pathophysiological role of cytokine interleukin-3 (IL-3) in early sepsis, and ii) a hybrid magneto-electrochemical sensor for IL-3 detection.

Materials & Methods: **Assay.** The IBS assay starts by extracting the target protein IL-3 through immunomagnetic enrichment. The captured protein is then subsequently labeled with detection antibodies and oxidizing enzyme (horseradish peroxidase, HRP). Finally, the beads are mixed with chromogenic electron mediators (3,3â€²,5,5â€²-tetramethylbenzidine, TMB); the enzyme catalyzes the oxidation of TMB and the reduction of H_2O_2. The oxidized TMB is then reduced by receiving electrons from the electrode, which generates electrical current as an analytical read out. Using magnetic beads allowed for direct and fast extraction of IL-3 from blood samples. Signal can also be amplified by concentrating magnetic beads underneath the electrodes. **Clinical study.** We applied IBS to detect IL-3 in clinical samples. Plasma or serum samples were collected from patients with symptoms of systemic infection, inflammation, or both. We obtained 23 samples from septic patients and another 39 from non-septic patients.

Results: The magneto-electrochemical sensing strategy developed here enables POC diagnostics: i) magnetic beads provides large surface area (~230 mm^2) for target capture and an easy way to perform assay steps; ii) the analytical signal is read out by compact, rugged electronics. Applying the IBS prototype, we achieved rapid (<1 hr), highly sensitive (<10 pg/mL) IL-3 detection in human plasma samples. Whole blood can also be used without preprocessing, which would better position the assay into most clinical workflows. The pilot clinical study supported the potential of IL-3 as a surrogate biomarker of sepsis: the sensitivity was 91.3%, and the specificity 82.4%. In comparison, procalcitonin (PCT) has been shown to have accuracies below 80%. Because IBS is fast and consumes small amount of samples (100 µL), it can be readily adopted to track temporal changes of biomarkers. This capacity would aid in not only the timely diagnosis of acute septic shock, but also reliable prognostication assessment of the risk. Also, using the small amount of samples would be especially advantageous for detection of sepsis in new borns/infants as blood samples from preterm infants (with a blood volume of <50 mL) are limited in volume.

Conclusion: The developed platform produces test results within 1 hour from native blood samples, and detects IL-3 at a sensitivity of <10 pg/mL; this performance is >5-times faster and >10-times more sensitive than a current gold standard, enzyme-linked immunoadsorbent assay. Using clinical samples, we show that high plasma IL-3 levels are associated with high organ failure rate and thus greater risk of mortality, confirming the potential of IL-3 as an early diagnostic biomarker. Compact and fast, the IBS platform can be readily integrated into clinical workflows, enabling timely diagnosis and proactive treatment of sepsis.

535063 Development of a Fast-Responding, Minimal-Equipment Biosensor for Zinc Deficiency

Abstract:

Zinc deficiency causes hundreds of thousands of childhood deaths annually, and current diagnostic tools are too expensive for widespread use in affected areas. Our group is working to meet the need for a low-cost diagnostic tool by developing a bacterial biosensor that produces different visible outputs based on the concentration of zinc in which the cells grow. To make this sensor easy-to-use and field-friendly, we have now engineered cells that completely repress pigmentation during growth, and then upon induction produce different pigments based on zinc concentration. This allows sample addition to concentrated pre-cultured colorless cells, allowing assay results in under 3 hours.

To do this, we engineered hybrid promoters that respond to both an exogenous inducer (to enable pigmentation) and zinc (to control which pigment is produced) by adding operator sites for the zinc-responsive repressor Zur to standard inducible promoters. The best-performing hybrid promoter was incorporated into circuits with other zinc-responsive elements to create a multicolor zinc sensor that is colorless during preculture and, upon induction, produces one of three pigments to indicate relative zinc concentration.

To increase clinical relevance, we next moved Zur's switching point from its natural threshold (0.1 μM zinc) to physiologically relevant serum zinc levels (5-10 μM). While analyte affinity of some sensors can be readily tuned, changing a transcription factor's affinity for its analyte by orders of magnitude without altering DNA binding activity is challenging. The most effective strategy to move the switching point was controlling Zur expression with a zinc-activated promoter. Through modulation of transcription factor expression levels, we created a library of sensors with switching points at extracellular zinc concentrations between 1 and 20 μM.

Taken together, our efforts have enabled a biosensor with potential for future impact on millions of lives in the developing world, with design and tuning strategies that may be generalizable to other nutrients and sensors.

517873 Non-Invasive Cell Density Measurement of Mammalian Cell Cultures in Early Stage Seed Trains

Abstract:

Mammalian cell culture processes rely on cell density to trigger process actions such as inoculum transfers. Currently, to determine cell density in the early stages of a cell culture seed train, shake flasks are opened, samples are removed, and the samples are tested on an automated cell density counting device. There are risks associated with frequent sampling and measuring cell density in shake flasks. These risks include potential contaminations from user operation and potential impact on cells as they are removed from the incubator. Multiple samples also reduce culture volume, which puts inoculation requirements for the next step at risk. A non-invasive cell density measurement for mammalian culture will greatly reduce risks of frequent sampling a culture. A handheld non-invasive biomass scanner, which was initially developed to obtain cell density for microbial cultures, was evaluated for its effectiveness in cell culture. In this study, the microbial biomass scanner was used to measure the total cell density in shake flask cultures of NS0 and CHO cell lines. The unitâ€™s effectiveness, ease of use, and accuracy will be presented and discussed.

525593 Protein Detection with Peptoid-Functionalized Carbon Nanotube Optical Sensors

Abstract:

A primary limitation to real-time imaging of small molecule metabolites and proteins has been the selective detection of biomolecules that have no naturally-occurring molecular recognition counterpart.[1] We present recent developments in the design of synthetic non-photobleaching fluorescent sensors, 'synthetic antibodies', to detect protein analytes, based on the near-infrared fluorescence modulation of single-walled carbon nanotubes (SWNT). Bio-mimetic peptoids, or N-substituted glycine polymers, are electrostatically pinned to the surface of SWNT to create peptoid-SWNT sensors sensitive and selective for the lectin protein wheat germ agglutinin.[2] We show the sensor to remain functional in Dulbecco's Modified Eagle's Medium, suitable for mammalian cell experimentation. The protein sensor is characterized with near-infrared spectroscopy and microscopy to extract protein-sensor interaction parameters and kinetic binding constants. We further show ternary selective interactions of the lectin with its conjugate sugars, through the near-infrared modulation of the peptoid-SWNT. This peptoid-SWNT sensor is a notable demonstration of a synthetic peptoid-nanoparticle biosensor with sugar recognition, and informs us of design considerations for developing synthetic molecular recognition elements to detect proteins and complex biomolecules.

1. Landry, M. P. *et al.* Single-molecule detection of protein efflux from microorganisms using fluorescent single-walled carbon nanotube sensor arrays. *Nat. Nanotechnol.* (2017). doi:10.1038/nnano.2016.284
2. Olivier, G. K. *et al.* Antibody-mimetic peptoid nanosheets for molecular recognition. *ACS Nano* **7**, 9276–9286 (2013).

538120 High-Throughput Single Cell Analysis of Deubiquitinating Enzyme Activity in Intact Cells

Abstract:

Molecularly-targeted therapeutics have shown significant promise in the treatment of several different cancers. In the treatment of multiple myeloma, drugs designed to inhibit members of the ubiquitin-proteasome system (UPS) have increased median patient survival rates. However, challenges in drug efficacy frequently occur due to tumor heterogeneity, drug resistance, and personalized responses. In the case of proteasome inhibition, recent studies have found that tandem therapies targeting both the proteasome and deubiquitinating enzymes (DUBs) have allowed cells to overcome drug resistance. DUBs are responsible for removing the polyubiquitin chain that targets proteins for proteasomal degradation. This work will demonstrate a new approach to directly quantify DUB activity and DUB inhibitor efficacy at the single cell level across a heterogeneous population. This is accomplished using a previously developed long-lived, cell permeable, fluorescent, DUB-specific, peptide-based reporter capable of delivering a visual readout of DUB activity in intact cells using fluorescent microscopy. Here, the peptide reporter is incorporated into a droplet microfluidic trapping array capable of encapsulating single cancer cells in aqueous droplets and trapping them in an overhead, circular array to facilitate direct imaging. The microfluidic platform is supplemented with a Python algorithm for rapid image analysis and hierarchical cluster analysis for single cell quantification of DUB activity and identification of distinct subpopulations with heterogeneous DUB activity. To assess difference in drug efficacy, the system was used to evaluate DUB activity in three different model myeloma cell lines challenged with different doses and combinations of DUB and proteasome inhibitors. This platform is a first of its kind to perform high-throughput single cell analysis of DUB activity and offers the potential for new studies in personalized diagnostics and drug development.

538597 A Handheld Optical Detection Method for Detection ofÂ *Vibrio Cholerae*Â in Environmental Water Samples

Abstract:

Environmental pathogen detection presents unique challenges in the development of novel biosensors due to the exceedingly low concentrations of pathogens in their native environments. For example, despite surviving at only 100 cells/mL in environmental water sources, the *Vibrio cholerae (V. cholerae)* bacteria that causes the devastating diarrheal disease cholera, leads to over 150,000 deaths worldwide each year. Further, the current gold standard for the detection of *V. cholerae* in water sources is an >8-hour process involving bacteria enrichment and culture followed by polymerase chain reaction (PCR). Despite being one of the most sensitive laboratory detection methods, PCR is not robust enough to directly detect *V. cholerae* from the environment; as it requires purification of DNA prior to amplification Hence, there is a need for a biosensor that can rapidly detect pathogens, such as *V. cholerae*, in their native matrix (environmental water sources).

In this work, we develop a highly accurate and sensitive biosensor for the rapid detection of *V. cholerae* in environmental water sources by pairing loop-mediated isothermal amplification (LAMP) with particlediffusometry (PD) (Figure 1A). We have developed a LAMP protocol for *V. cholerae* that is efficient (under 30-minute amplification), specific (targeting 6 different regions of the cholera toxin gene), and robust (usable in non-pretreated environmental water sources). When *V. cholerae* DNA is present in the solution, the LAMP assay polymerizes DNA targets into a variety of base pair lengths up to 25 kilobases. This polymerization causes an increase in the viscocity of the sample solution that can be measured via particle diffusometry (PD). In particle diffusometery we can calculate the change in the diffusion of nanoparticles in the sample using correlation-based algorithms of the particle images (Figure 1B). In this work, we show the applicability of PD-LAMP to detect the presence of *V. cholerae*. We demonstrate that PD-LAMP can be used to detect as few as 1 *V. cholerae* cells in a 25 μ L sample with only a 20-minute amplification period and an 8-second optical measurement time. Further, PD is a robust measurement system because it can detect amplified *V. cholerae* DNA from whole cells in environmental water without any additional enrichment or sample preparation steps. This detection method is 100-fold more sensitive than current gold standard fluorescence detection of nucleic acid amplification. We additionally demonstrate that detection of V. cholera from environmental water sources can be detected via PD using a novel handheld biosensor platform in as little as 30 min.

Figure 1. Illustration of PD-LAMP set-up. (A) The LAMP assay is performed in the presence of *V. cholerae* DNA (left). LAMP amplicons combined with polystyrene fluorescent particles (middle) are imaged under fluorescence microscopy (right). (B) Relationship of particle motion and viscosity. Particles undergo Brownian motion in a solution (left). In the presence of LAMP amplicons, the viscosity of the solution increases and particles experience hindered motion, indicating the presence of *V. cholerae* DNA in the sample (right).

540182 Phenome to Genome Enabled By Microfluidics and High-Throughput Quantitative Microscopy

Abstract:

Biological systems are complex. Often the observables are a result of gene and environmental interactions, sometimes influenced by stochasticity. To accelerate our understanding in order to cure diseases or to engineer solutions to industrial problems, mapping the relationships between phenome and genome is crucial. My lab is interested in developing and using a set of automation, microfluidics, and image-based data mining technologies to address questions in quantitative biology. I will show how we take advantage of simple hydrodynamics to design microfluidic systems for large throughput and spatially and temporally well-controlled experiments in Drosophila embryonic development as well as in immunology. In another example, I will show how we combine the power of experimental tools and computational tools to study problems in development neurobiology and behavior in C. elegans. The power of these engineered systems lies in that the throughput that can be achieved by using automation and microfluidics is 100-1,000 times that of conventional methods; furthermore, we can obtain information unattainable by conventional tools and â€œinvisibleâ€ to human vision to discover new genes and gene functions.

540575 Control of Pseudomonas Aeruginosa Biofilms By Electrical Currents Using a Simple Agar Model

Abstract:

Electroceutical wound dressings show promise for treating biofilm infections, however, the mechanism of action is poorly understood. We developed an in vitro agar based wound infection model using a bioluminescent strain of Pseudomonas aeruginosa to measure loss of activity and killing when direct current was applied to silver electrodes overlaid with agar. A 6V battery with 1kÎ© ballast resistor was used to expose the biofilms to the electric current for 1h and 24h. Loss of bioluminescence and a 4-log reduction in cells was achieved over the anode. Scanning electron microscopy showed that the biofilm had been disrupted and cells showed evidence of lysis. Even after turning off the current the antimicrobial activity continued to spread from the anode for 3 days suggesting generation of a stable antimicrobial agent had occurred. Elevated levels of chloride in the agar and chlorotyrosine in the dead biofilm suggested a role for hypochlorous acid. Similar killing was obtained for biofilms grown on gels containing onlybovine synovial fluid or human serum. We hypothesize that in the presence of physiological fluids, hypochlorous acid is produced at the anode which may react with proteins in the agar and/or the biofilm to form stable reactive intermediates.

524748 Prevention of Select Eskape Pathogens from Attaching to Titanium Using Cathodic Voltage Controlled Electrical Stimulation Combined with Antibiotic Therapy

Abstract:

INTRODUCTION: Infection following total joint arthroplasty (TJA) has been associated with longer patient healing times, prolonged hospital stays and increased costs to the healthcare system. ESKAPE pathogens are a concerning class of multidrug resistant bacteria in the medical setting that are contributing to these infections. In a previous study, we revealed that cathodic voltage-controlled electrical stimulation (CVCES) could prevent coupon-associated and planktonic colony forming units (CFU) of Gram-positive methicillin resistant *Staphylococcus aureus* and Gram-negative *Acinetobacter baumannii* from reaching detectable levels in a magnitude and time-dependent manner [1]. In this study we evaluated if application of CVCES at -1.0V, -1.5V, or -1.8V vs. Ag/AgCl, in conjunction with prophylactic antibiotics, is an effective method to prevent attachment of *Staphylococcus epidermidis* (*S. epidermidis*), *Klebsiella pneumoniae* (*K. pneumoniae*), *Acinetobacter baumannii* (*A. baumannii*), and *Pseudomonas aeruginosa* (*P. aeruginosa*) to the surface of commercially pure titanium (cpTi).

METHODS: cpTi coupons were wet sanded to a 600-grit finish, sonicated in deionized water for 10mins, and sterilized under ultraviolet light for 30mins. Experiments were conducted with blood isolates of Gram-positive *S. epidermidis*, strain SE94-309-0594, and Gram-negative *A. baumannii*, strain Ab307-029, *K. pneumoniae*, strain cKP1, and *P. aeruginosa*, ATCC(T) strain PA27853. Fresh cultures were inoculated into 5 mL of either tryptic soy broth supplemented with 0.25% glucose (TSBG, for SE94-309-0594) or Mueller-Hinton (MH) broth (Ab307-0294, cKP1, PA27853) at a 10:1 flask to volume ratio, adjusted to an OD_{600} of 0.1 and placed in a 37°C shaking water bath at 180 rpm. Cultures were grown to mid-log phase, diluted back to an OD_{600} of 0.1 and then further diluted 1:10,000 in the appropriate fresh media for a starting inoculum of approximately 10^3 CFU/mL. Minimum inhibitory concentrations (MICs) of antimicrobial agents were determined by the broth dilution method recommended by the Clinical and Laboratory Standards Institute (CLSI) [18]. The MIC for cefazolin against SE94-309-0594, Ab307-0294, and cKP1 was determined to be 1 µg/mL, 400 µg/mL, and 2 µg/mL, respectively. MIC of gentamicin was tested against PA27853 and found to be 3 µg/mL. Titanium was incubated for 24-hours at open circuit potential (OCP), OCP+drug, CVCES alone, or CVCES+drug. The stimulation chamber uses a standard three-electrode system with cpTi as the working electrode, a graphite counter electrode and an Ag/AgCl reference electrode [2]. With respect to Ag/AgCl, voltages of -1.0V, -1.5V, and/or -1.8V for 24 hours were applied to the working electrode at room temperature using a potentiostat (Interface 1000, Gamry Instruments). After incubation, cpTi coupons were washed in sterile PBS, sonicated and dilution plated for CFU. The media was collected for pH measurement and dilution plated for CFU. Four independent samples were evaluated in all conditions. A one-way ANOVA followed by Tukey post-hoc compared log-transformed CFU across stimulation conditions. If the Levene's test was significant, a Welch's test followed by a Games-Howell post-hoc test was performed. Student's t-tests were used to assess differences between independent stimulation conditions with and without antibiotic treatment.

RESULTS: Remarkably, CVCES of *P. aeruginosa* at -1.0V+gentamicin and -1.5V+gentamicin showed an 8-log reduction, to below detection, from controls and CVCES alone. -1.0V+cefazolin significantly reduced coupon-associated *A. baumannii* from control ($p<0.001$), and -1.5V resulted in no detectable coupon-associated CFU and reduced planktonic CFU to below 10^1 CFU/mL. Compared to controls -1.8V and -1.8V+cefazolin prevented all *K. pneumoniae* from reaching detectable levels ($p<0.001$), but had no effect at -1.5V and -1.5V+cefazolin. -1.0V+cefazolin to *S. epidermidis* reduced planktonic CFU by 93.2% and resulted in no detectable coupon-associated CFU. -1.5V alone significantly reduced *S, epidermidis* coupon-associated and planktonic CFU from controls ($p<0.005$), and prevented all CFU at -1.5V+cefazolin (data not shown). Increasingly cathodic CVCES had a higher average cathodic current density and cumulative charge transfer. CVCES can prevent coupon-associated and planktonic CFU of select ESKAPE pathogens from reaching detectable levels on cpTi, and antibiotics synergistically enhance these effects.

DISCUSSION: Implant associated infection following orthopedic intervention is a devastating clinical complication associated with increased patient morbidity, extended treatment times, and increased costs to the health care system. The purpose of this *in vitro* study was to evaluate if CVCES of titanium in conjunction with antibiotics could synergistically prevent initial attachment and/or reduce the viability and subsequent bacterial biofilm formation of select ESKAPE pathogens. The outcomes of this study revealed that CVCES could prevent coupon-associated CFU and planktonic CFU of select ESKAPE pathogens from reaching detectable levels in a magnitude-dependent manner. In general, increasingly cathodic potentials had more robust antimicrobial effects and this effect was enhanced in the presence of antibiotic treatment. Bactericidal activity of stimulation corresponded with an alkaline shift in pH, which is thought to be the causative link between stimulation and bacterial cell death. In general, the addition of clinically relevant antibiotics enhanced the efficacy of CVCES to have a more robust preventative effect and these effects were dependent on the magnitude, antibiotic, and species being tested.

SIGNIFICANCE: Current antibiotic prophylactic treatment methods for joint revision surgeries are not effective against all pathogens; however alternative methods, such as CVCES in conjunction with antibiotics, shows promise to be an effective broad-spectrum prevention method for Gram-positive and Gram-negative pathogens.

REFERENCES: [1] Canty M, *et al.* Acta Biomaterialia: 2017;48:451-460 [2] Ehrensberger MT *et al.* Biomaterials 2015;41:97-105.

537838 Computational Modeling of Cathodic Voltage Controlled Electrochemical Treatment of Biofilms in-Vivo

Abstract:

Computational Modeling of Cathodic Voltage Controlled Electrochemical Treatment of Biofilms In-vivo

Amir Mokhtare[1], Mark Ehrensberger[2]and Edward P Furlani[1,3]

[1]Dept. of Chemical and Biological Engineering,[2] Dept. of Biomedical Engineering, [2] Dept. of Biomedical Engineering, [3]

University at Buffalo, The State University of New York, Buffalo NY

Email: efurlani@buffalo.edu Abstract

The last few decades have witnessed a proliferation in the use of prosthetic implants. Chronic bone and joint related disorders such as osteomyelitis, septic arthritis, etc. are widespread and can have a devastating impact on quality of life. Consequently, the use of total joint arthoplasty (TJA) has grown steadily as a life enhancing procedure. Although successful joint replacement restores functionality and provides pain relief, prosthetic joint infections (PJIs) can be a devastating outcome of orthopedic surgery and can impose horrendous physical burdens on individual patients and financial stress on the health industry as a whole. PJIs are often caused by aggregated communities of bacteria known as bacterial biofilms, which tend to be incredibly resistant to antibiotics. As a result, new methodologies are needed to efficiently treat biofilm-based PJIs in-vivo. In this regard, electrochemical treatment of the infected implants is among the most promising strategies for the eradication of surface-based biofilm infections. Cathodic voltage controlled electrical stimulation (CVCES), is a recently developed electrochemical approach that has proven to be effective in the treatment of biofilm-based PJIs.

Analysis of in-vivo assays have revealed that CVCES combined with conventional antibiotic drugs can lead to complete treatment of the infections without any histological damage to the surrounding tissue. However, many fundamental aspects of this treatment are unknown and rational design towards optimization is lacking. In this presentation we introduce a one-dimensional (1D) computational modeling framework to simulate the electrochemical behavior of an in-vivo CVCES system. We use dilute solution theory (ionic interaction are neglected) for determining the electrochemical potential of ionic species in the system and solve the Nernst-Plank equations along with Maxwell's equations to predict the toxic (to bacteria) species concentration speciation and electric potential. Water-based electroneutrality is used as a simplifying assumption to facilitate numerical convergence. In addition, Butler-Volmer electrochemical surface reaction kinetics are assumed at the surface of the working electrode and as boundary conditions for the models. We have used our experimental current-voltage scanning data to identify the dominant electrochemical surface reactions and have extracted transfer coefficients and exchange current densities required to characterize the reactions. The surrounding tissue, i.e. the electrolyte medium in the in-vivo experiments, was approximated by the dilute solution of sodium chloride. Moreover, the most important buffer systems inside the tissue, namely the bicarbonate buffer system and buffer capacity of proteins and organic phosphates, were introduced in the form of homogeneous reactions. Hydrogen evolution and oxygen reduction electrochemical reactions are assumed at the working electrode surfaces and characterized by voltage current scanning data. Both experimental and mathematical modeling results showed that oxygen reduction surface reaction become diffusion limited after few minutes and only contribute a small amount to the total current density at the electrode surface. Simulation results also elucidate how buffering capacity of the tissue would counteract the diffusion of hydroxide ions in the tissue. The mathematical model also provides the temporal-spatial variation of the both current-density and pH profiles that are very difficult to obtain in experiments. The simulation results are in an acceptable quantitative range with the experimental results. The modeling framework can be extended to higher dimensions to represent more realistic examples such as human knee or hip implants. Such models can be utilized to optimize the electrode configurations and operation parameters.

Keywords: Biofilm induced infections, prosthetic joint replacement, in-vivo mathematical modeling, electrochemical treatment. Results

Figure1. Schematic of the in-vivo assay and computational domain, a) pH profile distribution

b) concentration profiles c) IV experimental and modeled data d) transfer coefficients of hydrogen evolution reaction

References

[1] M. T. Ehrensberger *et al.*, *Biomaterials,* vol. 41, pp. 97, 02/ 2015.

[2] F. L. Hellweger *et al.*, *Nature Reviews Microbiology,* vol. 14, no. 7, pp. 461, 2016.

[3] E. Nilsson *et al.*, *Bioelectrochemistry,* vol. 53, no. 2, pp. 213, 2001.

[4] S. Nodzo *et al.*, *Clinical Orthopaedics and Related Research®,* vol. 473, no. 9, pp. 2856, 2015// 2015.

538388 Electroactive Surfaces and Their Use for Biofilm Removal to Advance Wound Healing

Abstract:

Biofilm infections are present in nearly all chronic wounds, contributing to delayed wound healing. Biofilms are difficult to treat due to their dense structure, slow growth rate, altered metabolism, and protection by extracellular polymeric substances. Current approaches to biofilm infection treatment use a combination of debridement, and systemic and topical antibiotics. These approaches partially treat biofilm infections. However, their performance is hindered due to the depletion of active compounds at deep layers of biofilms and the relative non-susceptibility of biofilm bacteria to conventional antibiotics. This gives rise to persister cells that survive treatment, and leads to infection recurrence. In this work, we present a novel electroactive material surface and an electronic platform for the management of wound biofilm infections by continuously producing effective concentrations of antimicrobials near the electroactive material surface. The concentration of antimicrobials is tuned using a miniaturized electronic device so that the concentration is sufficient to eliminate biofilm infections without apparent damage to host tissue. The efficacy of this strategy was evaluated for the treatment of common biofilm forming pathogens using *in vitro* and *ex vivo* biofilm models. We used mathematical modeling with finite-element analysis software COMSOL Multiphysics® to predict the antimicrobial generation rate near electroactive material surfaces.

539225 Toward to the Design of an Electrochemical Therapy (ECT) Against Microbial Infection

Abstract:

One of the main challenges healthcare faces is the preponderance of antibiotic-resistant microorganisms in public and medical environments. The antibiotic-resistant species constitute significant challenges to curing chronic infections, especially those associated with implanted medical devices. We explore the antimicrobial activity of low-level electrochemical treatments on a range of microorganisms. We report that these pathogenic microbes can be effectively eliminated by low-level direct currents (DCs); e.g. treatment with 70 Î¼A/cm^2 DC for 1 hr using titanium or stainless steel (SS) 304 electrodes reduced the number of viable planktonic *Streptococcus mutans*, *Candida albicans*, *Pseudomonas aeruginosa* by 1-2 logs, compared to the untreated control. In addition to killing by applying DC alone, synergistic effects are achieved when treating the planktonic and biofilm cells with 70 Î¼A/cm^2 DC and antibiotic dosages concurrently. Interestingly, alternative ECT utilizing DC generated with graphite followed sequentially by 1.5 Î¼g/mL Tob, or DC via TGON 805 (a carbon-based material) as a sole factor, eradicated the persisters completely (~7-log). Electrochemical factors including electrode composition, electrochemical byproducts, and current/potential levels contributed to impairment of persistersâ€™ electrophysiology, causing substantial changes in cell structures and transcriptomes. This work will define the conditions necessary to modulate electrochemically-induced sensitization of pathogens and eradicate polyspecies biofilms from implantable materials.

530184 Wireless Electrostimulation to Eradicate Bacteria Biofilm

Abstract:

Pathogenic bacteria such as *Pseudomonas aeruginosa* and *Staphylococcus aureus* can form biofilms on medical implants and cause serious infections that are incurable by conventional antibiotics due to high-level drug resistance. As an alternative approach, electric currents have been shown to kill biofilm cells and have synergy with antibiotics in bacteria killing. However, conventional electrochemical treatment requires connection electrodes to a power source. This requires invasive procedures and can cause complications, hindering the application of this technology *in vivo*. In this study, we demonstrate that biofilm cells can be effectively killed by using electromagnetic induction to deliver direct current wirelessly from a remote power source. For example, after treatment with 60 $\mu A/cm^2$ of wirelessly delivered DC conducted by stainless steel electrodes for 6 hours, the viability of biofilm cells was reduced by approximately 4 logs and 2.2 logs for *P. aeruginosa* and *S. aureus*, respectively. Similarly, the viability of *P. aeruginosa* and *S. aureus* biofilm cells was reduced by 2 and 2.5 logs after treatment with the 30 $\mu A/cm^2$ of wirelessly delivered DC conducted by graphite-based electrodes for 3 hours, respectively. Synergy in concurrent treatment with lower level DC and antibiotics to biofilms was also observed. These conditions appeared to be safe to the human epithelial cells. Encouraged by these findings, we engineered a model that contains the wireless DC treatment units and shares similar profile of implantable medical devices. The device was found to effectively kill of *P. aeruginosa* and *S. aureus* biofilms without noticeable toxicity to human cells. With the capability to kill bacteria without using a directly connected power source, this platform has potential applications in engineering smart devices to non-invasively eradicate biofilm infections.

527193 Electrochemical Detection of Bacterial Biofilms on Titanium

Abstract:

Electrochemical Detection of Bacterial Biofilms on Titanium

Caelen M. Clark and Mark T. Ehrensberger

State University of New York at Buffalo, Buffalo NY, USA

Statement of Purpose: Periprosthetic joint infection (PJI) is a devastating complication of total joint arthroplasty. In addition to being extremely difficult to treat with antibiotics alone, PJI is also difficult to detect [1]. These problems stem from the ability of bacteria to form biofilms on the surface of implanted materials. When in the biofilm state, bacteria gain a diffusional barrier which limits the penetration of antibiotics, and causes a reduction in the metabolic activity of the resident cells. This can make detecting biofilms with traditional culture methods difficult. For this reason, new methods for the detection of biofilms on implanted orthopedic materials are needed. This work evaluated the electrochemical methods of potentiometry and electrochemical impedance spectroscopy (EIS) as diagnostic measures of bacterial biofilm formation on titanium in an *in vitro* model.

Methods: A custom designed polycarbonate electrochemical biofilm reactor (Fig 1) was utilized to both grow biofilms and make the electrochemical measurements.

Figure 1: Experimental setup of the biofilm reactor

Samples were made from grade 4 commercially pure titanium (Ti). A clinical isolate of Acinetobacter baumannii (strain Ab307), a problematic Gram-negative human pathogen, was used in all experiments. Biofilms were grown by introducing a 30 mL inoculum with a concentration of $\sim 1\times 10^7$ CFU/mL to the reactor. After two hours, flow in the system was initiated at 0.02 mL/s. The reactor also functioned as a three electrode electrochemical cell utilizing the Ti as a working electrode, an Ag/AgCl reference electrode, and a graphite counter electrode. The open circuit potential (OCP) of the Ti was measured prior to inoculation of the reactor, and 24 and 48 hours after inoculation. EIS was performed using a Gamry Interface 1000 potentiostat with a 10 mV sinusoidal voltage about the OCP in the frequency range of 100 kHz-5 mHz at the same timepoints as the OCP measurements. The results of the EIS spectra were fit to a modified Randles circuit, and the polarization resistance (Rp), constant phase element (CPE) magnitude (Y0), and CPE exponent (α) were calculated. Following EIS, the Ti coupons were extracted, rinsed in phosphate buffered saline, sonicated in 0.1% saponin, and dilution plated for enumeration of CFU. Scanning electron microscopy (SEM) of the extracted Ti was also performed for qualitative assessment of bacteria on the Ti surface. Control experiments were carried out with the same experimental conditions without bacterial inoculation. A total of 9 samples used for both the biofilm and the control experiments, and a T-test was used to compare the electrochemical parameters at each time point.

Results: Biofilm formation on Ti resulted in a significant cathodic shift in the OCP after 24 hours as compared to the control samples. The Rp was found to increase significantly after 24 hours of biofilm growth (Table 1). CFU enumeration and SEM imaging (Fig 2) confirm the presence of biofilms with an average of 1.1×10^7 CFU/mL.

	OCP (mV)		
Time	**Ab307**	**Control**	**P-Value**
0	-326±32		
24	-387±23	-321±20	<0.001
48	-381±21	-308±32	<0.001
	Rp (Ω/cm^2)		
	Ab307	**Control**	**P-Value**
0	$6.98\times10^6 \pm 4.6\times10^6$		
24	$5.4\times10^7 \pm 2.8x10^7$	$1.0x10^7 \pm 6.0x10^6$	<0.001
48	$1.1\times10^8 \pm 1.1x10^8$	$1.8x10^7 \pm 2.8x10^7$	0.04

Table 1: Electrochemical measurement results

Figure 2: SEM micrograph of Ab307 biofilm

Conclusions: Biofilms grown on Ti for 48 hours resulted in a statistically significant cathodic shift in the OCP of the metal at 24 and 48 hours compared to control samples. The Rp was also found to increase significantly compared to control samples after 24 hour of biofilm growth. These results show the effect of biofilm formation on the electrochemical properties of Ti. Further work will be conducted to track these electrochemical parameters as a function of biofilm formation and growth.

References: [1] Sendi P, Zimmerli W. International Journal of Artificial Organs. 2012;35:913-22.

534597 Novel Focused Multivector Ultraviolet (FMUV) Disinfection without Manual Cleaning and Chemical Disinfection in-between Surgeries and throughout the Hospital Environment

Abstract:

PurpleSun is an early stage healthcare technology company based out of New York City. We are focused on providing smart, light-based technology solutions to the healthcare industry for infection prevention, diagnostics and treatment.

Our companyâ€™s novel light technology platform will be utilized throughout the continuum of care, within hospitals, long term care and surgical centers. The smart devices will set the stage for patient safety through the highest of quality in medical equipment disinfection.

Our goal is to prevent healthcare acquired infections, which will ultimately save lives, reduce costs and enhance safety for patients and caregivers. Our clinical data proves best in class efficacy and performance of our platform, and positions PurpleSun to illuminate the healthcare arena.

534615 Boronate Ester-Based Dynamic Nucleic Acids for Templated Analyte Detection

Abstract:

Beyond storing genetic information, nucleic acids have become incredibly important tools for self-assembly, nanotechnology, drug delivery, and biosensing applications because of their unique molecular recognition properties. For example, in addition to recognizing complementary nucleic acid sequences, DNA and RNA have been used as molecular recognition agents called aptamers, which detect a variety of ligands, ranging from small molecules to proteins. Aptamers are generated for specific ligands by using systematic evolution of ligands by exponential enrichment (SELEX), in which the desired ligand is incubated with a large library of oligonucleotides and through repeated cycles of amplification, the oligonucleotides with the highest affinity for the ligand are selected. Alternatively, researchers have begun exploring the use of dynamic covalent chemistry to assemble synthetic nucleic acids in the presence of template molecules.

In this work, we develop a system that can be used for templated-assembly of synthetic nucleic acids based on dynamic boronate esters. The boronic acid-containing polymer was synthesized via a copolymerization of maleic anhydride and 4-vinylphenyl boronic acid. Maleic anhydride was included as a comonomer to improve water solubility and to achieve regular spacing of the boronic acid functionalities. Polymerization conditions, specifically solvent, temperature, and comonomer ratio, were varied in order to achieve an alternating structure. Apparent dissociation constants for four ribonucleosides (i.e., 5-methyluridine, cytidine, guanosine, and adenosine) were determined via a colorimetric, competitive binding assay with Alizarin Red S. The influence of buffer and DMSO content on ribonucleoside binding was investigated. By identifying conditions in which all four ribonucleosides bind to the polymer with similar affinity, template-directed incorporation of the ribonucleosides can be achieved.

523638 Mechanistic Optimization of Floating Gate Transistors for Biosensing Applications

Abstract:

Transistors have been utilized extensively as biosensing transducers due to their fast and sensitive signal transduction. The Electrolyte-Gated Transistor (EGT), in which the conventional gate dielectric is replaced with an electrolyte, is particularly attractive because it offers the added advantage of low voltage operation, and can be made using manufacturing techniques, such as printing, that potentially lower cost. The Floating Gate Electrolyte-Gated Transistor (FG-EGT, or FGT) is a device which incorporates a floating gate that physically separates and electronically couples the active sensing area with the transistor. This eliminates degradation and optimization issues associated with interactions of the semiconductor/dielectric with the analyte medium, and allows for independent choice of electronic materials and capture chemistry. The floating gate therefore allows it to overcome several shortcomings of traditional transistor based sensors that utilize the semiconductor or dielectric surface for molecular capture. Previous work has demonstrated the utility of FGTs for the fast and reliable detection of ssDNA in buffers, ricin in potable liquids, and gluten from different grain sources.

The detection mechanism of FGTs has been attributed previously to surface charge and capacitance changes at the sensing surface. In this work, we demonstrate charge vs. capacitance based optimization of the sensing pad to provide maximum sensitivity for different types of targets. Self-assembled monolayers of MUA and alkanethiols were utilized to systematically alter the charge state and capacitance of the sensing surface, and the results obtained allowed us to optimize the sensing area with regard to both the charge and size of target molecules. We further optimized the microfluidic channels used for target delivery in terms of height and aspect ratio to increase target molecule flux at the surface and enhance detection times for the device. Our study provides general conclusions regarding testing protocols, sensing areas, and target delivery methods that are applicable to a wide range of transistor based biosensors, which are a promising class of sensing devices for food safety and point of care diagnostics.

530403 Computational Optimization of Metal-Organic Framework (MOF) Arrays for Chemical Sensing

Abstract:

Electronic noses are used in applications ranging from food quality assessment to environmental monitoring. However, current sensor technology is inadequate to properly address todayâ€™s challenges, such as low-concentration methane detection and disease detection via breath analysis. Metal-organic frameworks (MOFs), having high surface areas and tunable pore structures are ideal candidates as substrates for improved gas sensors. We use computational screening to predict the adsorption of gas mixtures of N_2, CH_4, and O_2 onto arrays of MOFs to identify the materials that maximize the signal-to-noise ratio for gas detection, obtaining sensor response signals from this adsorption data. Subsequently, an algorithm employs statistical methods to estimate the component concentrations in unknown gas mixtures probabilistically. Additionally, we quantify the information content of each possible MOF array, calculated via the Kullback-Liebler Divergence. Ultimately, we show how this computational approach is dramatically more efficient than choosing sensing materials for electronic noses by trial-and-error. For example, given 9 MOF materials to construct a 4-element array, rather than building and testing all 126 arrays, the best one can be computationally predicted in minutes. Such improvements will expedite the development of comprehensive electronic noses for hazardous materials sensing and disease detection.

531807 Morpholino Materials for Diagnostic Applications

Abstract:

Biosensing applications rely on careful tuning of molecular properties to optimize the specificity, sensitivity, and speed of an assay. Since biomolecules are typically charged, such optimization often involves consideration of charge effects. We have been examining the impact of charge in the context of nucleic acid assays by comparing the performance of DNA and morpholino monolayers for detection of complementary nucleic acid sequences. Morpholinos are synthetic DNA analogs based on an uncharged backbone. This talk will describe prospective benefits of morpholinos' nonionic character for diagnostic applications, with emphasis on electrostatic control of the underlying hybridization reactions. For example, in contrast to DNA probes, morpholinos enable facile cycling between hybridized and dehybridized states. The impact of ionic strength and temperature on effectiveness of electrostatic control is evaluated, as is the impact of mismatches. Optimal control is found when stability of probe-analyte duplexes is tuned so that electrostatics can efficiently switch between the forward (hybridization) and reverse (dehybridization) directions.

527513 Cardiac Troponin I Detection Using Antibody-Immobilized Disposable Cover Glass and AlGaN/GaN High Electron Mobility Transistors

Abstract:

Cardiac troponin I (cTnI) detection was demonstrated using a functionalized disposable cover glass connected to an AlGaN/GaN high electron mobility transistors (HEMT). Cardiovascular disease is one of the leading causes of mortality in the US, and it has caused tremendous hardship for both individual health and finance in United States. Every year, over 800,000 Americans die due to the cardiovascular diseases.[1] Troponins are released when the cardiac and skeletal muscles are damaged. There are three types of troponins; troponin I, troponin T, and troponin C . Troponin I is extremely cardio-specific, and the elevation of troponin I concentration in the blood stream indicates the cardiac muscle damage. Troponin T is also a biomarker for myocardial infarction patient, however it is not exclusive. Studies have shown the elevated troponin T were also detected for patients with muscular dystrophy and renal disease. With the non-cardio specific nature of troponin C, it has never been used for clinical practice as a myocardial necrosis marker. [2,3,4] Currently, all the detection methods and systems for cTnI are time consuming, expensive and require a well-trained technician to perform and complete the tests.

The challenge for cardiac Troponin I (cTnI) detection method in clinical practice is to develop a hand-held, low cost, accurate, and real-time biosensors for myocardial infarction patients. The combination of functionalized cover glass and AlGaN/GaN high electron mobility transistors (HEMT) detection method has overcome the primary difficulty of biomarker solution with high ionic strength in traditional HEMT biosensors. In this approach, the HEMT device is not exposed to the bio-solution, which can be used repeatedly. The sensor functionalization was performed on a low cost disposable glass slide, which can be viewed as testing strips. Furthermore, the cover glass could be replaced by plastic or paper for cTnI detection, which can be more cost-effective. To test the cTnI concentration, pulsed bias-voltages are applied to a functionalized electrode on the cover glass and the drain terminal of the HEMT device. The electrical pulse would result in the formation of the electrical double layers (EDL) on the active sensing area, and different cTnI concentration would lead to different charge potential that is applied onto the conducting channel of this HEMT device. The detected cTnI concentration range was achieved from 0.1 ng/ml to 100 ng/ml. This covers wide range of cTnI concentration for the myocardial infraction, such as the elevated cTnI concentration of 10ng/ml or higher for acute myocardial infraction (AMI), and 0.5 ng/ml to 2.0 ng/ml cTnI concentration for early stage of AMI patients. The static drain current changes as a function of cTnI concentrations were modeled with the Langmuir extension model. Also, the spring relaxation model was employed for the dynamic modeling of the time-dependent current response for different biomarker concentrations. The results in this study demonstrated that the functionalized cover glass approach is a very attractive cTnI detection technology for inexpensive cartridge-type heart attack biosensors, and this technology can be applied to other sensing areas such as cancer detection, environmental monitoring, and food safety.

1. Benjamin, E.J., Blaha, M.J., Chiuve, S.E., Cushman, M., Das, S.R., Deo, R., de Ferranti, S.D., Floyd, J., Fornage, M., Gillespie, C., Isasi, C.R., JimÃ©nez, M.C., Jordan, L.C., Judd, S.E., Lackland, D., Lichtman, J.H., Lisabeth, L., Liu, S., Longenecker, C.T., Mackey, R.H., Matsushita, K., Mozaffarian, D., Mussolino, M.E., Nasir, K., Neumar, R.W., Palaniappan, L., Pandey, D.K., Thiagarajan, R.R., Reeves, M.J., Ritchey, M., Rodriguez, C.J., Roth, G.A., Rosamond, W.D., Sasson, C., Towfighi, A., Tsao, C.W., Turner, M.B., Virani, S.S., Voeks, J.H., Willey, J.Z., Wilkins, J.T., Wu, J.H., Alger, H.M., Wong, S.S., Muntner, P., 2017. Heart disease and stroke statisticsâ€"2017 update: a report From the American Heart Association. Circulation (2017)
2. Mythili and N. Malathi, Diagnostic markers of acute myocardial infarction. Biomed. Rep. 3, 743 (2015).
3. Amitava Dasgupta and Amer Wahed, Chapter 8 - Cardiac Markers, In Clinical Chemistry, Immunology and Laboratory Quality Control, Elsevier, San Diego, 2014, Pages 127-144, ISBN 9780124078215
4. Allan S. Jaffe, Vlad C. Vasile, Margherita Milone, Amy K. Saenger, Kalen N. Olson, Fred S. Apple, Diseased Skeletal Muscle: A Noncardiac Source of Increased Circulating Concentrations of Cardiac Troponin T, Journal of the American College of Cardiology, Volume 58, Issue 17, Pages 1819-1824 (2011)

537053 Nanostructured Polymeric Membranes for in-Situ Measurement of Exhaled Formaldehyde and Acetone Kinetics As Early-Stage Non-Invasive Markers of Lung Disease

Abstract:

Early stage discovery of lung cancer can significantly improve patient prognosis. Unfortunately, the impact of a well-known technique such as computed tomography screening on mortality rates is uncertain. It has been postulated that tumor growth may lead to peroxidation of cell membranes and is responsible for the observed emission of volatile organic compounds (VOCâ€™s) such as acetone and formaldehyde. This observation has stimulated significant recent effort in the measurement of these compounds in exhaled breath. While promising approaches have been developed for analyzing average VOC concentrations in exhaled breath, an unrecognized drawback to current efforts is the presumption that gas partitioning between the blood and gas phases occurs so rapidly that the time averaged concentration of collected VOCs is representative of their saturation concentration in the lungs. In fact, when one considers the mass transport resistances encountered, VOC lung concentration is likely to be time-dependent and any sensing approach must have sufficient time resolution to capture the dynamic evolution of this concentration during a typical breath exhalation period of not more than 40 seconds in a high humidity environment.

In this presentation, we describe development of an optical exhaled gas sensing approach utilizing a nanostructured polymeric membrane that meets these requirements and demonstrates the dynamic evolution of exhaled breath biomarker concentration. We show that the activity of a solid acid catalyst toward the heterogeneous condensation reactions of immobilized resorcinol reagent with gas-phase acetone and formaldehyde can be preserved even at 100% ambient relative humidity through the incorporation of organic acids such as vanillic or tiglic. The reaction produces a colored flavan and poly-condensation products that permits highly selective and sensitive correlation to acetone and formaldehyde concentrations in exhaled breath. Such behavior relies on the complex hetero-phase morphology of the membrane as will be described.

528113 Invited Talk: Advancing Biosensing with Hybrid Nanomaterials and Machine Learning

Abstract:

Single-walled carbon nanotubes (SWCNTs) and more recently graphene derivatives have attracted considerable interest for the development of chemical and biological sensors. These carbon nanostructures are just one atom thick, and their electronic properties are extremely sensitive to adsorption of chemical species on their surface. When decorated with metal or metal oxide nanoparticles, these nanostructures exhibit a large and selective electronic response toward many analytes with potential applications in medical diagnostics. However, achieving selectivity through chemical functionalization necessitates destroying the sp^2structure of SWCNTs, leading to a partial or total loss of electrical properties. To circumvent this issue, holey graphene may be used to accomplish covalent chemistry via coupling with graphene edges through carboxylic acid moieties. While pristine graphene is a zero-bandgap semiconductor, holey graphene mimics graphene nanoribbons, producing a bandgap at neck widths less than 10 nm. The resulting structure is analogous to an interconnected network of SWCNTs on a device surface and holds promise for chemical and biological sensing.

Sensor array approaches involving multivariate data analysis and ultimately machine learning techniques to classify different analytes have many emerging applications. For example, sensing specific biomarkers for cells is very challenging. The specific biosensing methods often require foreknowledge of the biomarker in question to successfully identify its presence. For biosensors, this necessitates individually designed devices for each specific target of interest. Carbon nanotube-based field-effect transistors (NTFET) are highly sensitive to changes in the local environment and are an ideal platform for the fabrication of label-free biosensors. Characteristic FET curves contain a wealth of information and have proved useful for analyzing the mechanism of interaction with target analytes. This rich information provides an interesting target for deeper analysis towards more general biosensing applications.

536264 Viscosity Measurement at the Point-of-Need: A Smartphone Capillary-Based Approach

Abstract:

Viscosity, an essential property of fluid behavior, plays an important role in many applications including quality control in chemical, food, and manufacturing industries, as an analytical tool in research and development settings, and in medical diagnosis, prognosis, and preventive medicine. Conventional viscometers typically require large sample volumes (several mL), use non-disposable sample holders, involve time-consuming protocols, require trained operators, lack portability, and are expensive. These limitations have hindered widespread use of viscosity measurements at the point-of-need, especially important for medical applications. Blood viscosity, for instance, has been shown to be a predictor of cardiovascular events in hypertensive men and it could be a risk factor for diabetes. Here we present a simple smartphone-enabled capillary-based approach for the measurement of a liquidâ€™s viscosity using small sample volumes (~0.03 mL). A smartphone is employed to record short videos (at up to 240 frames per second) of capillary-driven flow in disposable glass capillaries of uniform circular cross-sections (100-800 micrometers in diameter) in a horizontal configuration. Videos of capillary filling, which typically lasts anywhere from 0.5 to 10 seconds, are then processed using automated video analysis to detect fluid displacement and retrieve the meniscus position as a function of time. Further analysis of the retrieved data provides rheological properties of the fluid while including the effects of a dynamic contact angle observed in capillaries with larger diameter. Results from our viscometer are in quantitative agreement with known fluids used as references, such as Newtonian water-glycerol solutions and shear-thinning water-xanthan gum solutions, while blood viscosity measurements illustrate its biosensing capabilities. The materials needed (capillaries and a smartphone camera) and the analysis performed (implemented as a user-friendly software) enable viscosity measurements to be performed rapidly and without requiring special knowledge or training. Thus, this approach is low cost and portable, overcomes important limitations of conventional viscometers, and facilitates measuring the rheology of various fluids in point-of-use and resource-limited settings.

536391 Corona Phase Molecular Recognition Sensors in Marine Organisms for Physiological Biologging: A Feasibility Study

Abstract:

For decades, biologgers have been tagging animals with various types of sensors to study aspects of their behavior unable to be queries in controlled laboratory environments. To date, environmental and physiological chemical information remains an underutilized data source that offers tremendous potential to raise the level of insight gained from such studies. In this work, we explored the feasibility of applying a synthetic chemical sensing technology, Corona Phase Molecular Recognition (CoPhMoRe), to marine organism biologging. Fluorescent DNA-wrapped single wall carbon nanotubes (SWNT) were fabricated, encapsulated in a biocompatible hydrogel, and shown to experience fluorescence quenching when exposed to riboflavin, a key nutrient in oxidative phosphorylation. Over three days at Oceanogràfic in Valencia, Spain, animal experiments were performed in seven species of fish, sharks, eels, and turtles. Hydrogels were delivered subcutaneously via trocar and imaged using an inexpensive Raspberry Pi camera system. Hydrogels could be detected up to a depth of 7 mm in teleost and catshark species. The effects of tissue heterogeneity on hydrogel delivery and fluorescence visibility were also explored, with darker tissues masking hydrogel fluorescence. Hydrogels were implanted into a living turtle, eel, and catshark. The catshark and eel ls displayed no adverse health effects and or changes in tissue architecture 2 months after the hydrogel implantation, indicating good tolerance of the implant. The turtle experienced a foreign body reaction to the implant. All three animals displayed similar movement and feeding behaviors before and after the implant. Taken together, these results indicate that the application of CoPhMoRe to marine organism biologging is indeed feasible and warrants future development. Future studies will refine the sensor configuration and use a larger numbers of animals.

534886 Detection of Beta Carotene and Lutein Using Electrochemical Impedance Spectroscopy

Abstract:

Carotenoids are biological pigment molecules that are often found in fruits, vegetables, and human tissue. Humans cannot synthesize carotenoids, but they can be accumulated through diet [1]. Carotenoids have been found to lower oxidative stress which can lead to a lower risk of disease and infections [2]. Lower carotenoid levels have also been correlated to patients with Alzheimer's Disease, Parkinson's Disease, and multiple types of cancer [3,4]. Although there is a reverse correlation between the concentration of carotenoids and risk of disease, there are still a lot of qualities of carotenoids that are unknown to scientists. An effective point-of-care, cheap, and sensitive carotenoid sensor would aid in showing causality in human patients using carotenoid levels as a potential screening biomarker.

Electrochemistry is a cost effective sensing paradigm while maintaining sensitivity [5]. Electrochemical impedance spectroscopy (EIS), an electrochemical method that measures the surface impedance when an alternating current is applied, is known to be an extremely sensitive surface technique. We investigated the use of EIS to correlate carotenoid solution concentration to glassy carbon surface interactions. Nanomolar levels of β-carotene and lutein were successfully detected using EIS in a solution with only analyte and in the presence of each other. Uniquely reported here will be the sensitivity and selectivity of β-carotene in the presence of lutein and the lutein in the presence of β-carotene. Surface-bound β-cyclodextrin was used to increase carotenoid detection sensitivity. β-cyclodextrin was chosen because of its ability to create host-guest complexes with hydrophobic molecules such as β-carotene and lutein. The sensitivity of carotenoid detection using β-cyclodextrin will be reported compared to unmodified electrodes. In conclusion, the significant progress we have made over the past year (i.e. detecting carotenoids in the presence of competing molecules and cyclodextrin surface modifications) will be reported and discussed.

[1] J. Fiedor, K. Burda, Potential Role of Carotenoids as Antioxidants in Human Health and Disease, Nutrients. 6 (2014) 466–488. doi:10.3390/nu6020466.

[2] J.A. Mares-perlman, A.E. Millen, T.L. Ficek, S.E. Hankinson, Symposium : Can Lutein Protect Against Chronic Disease ? The Body of Evidence to Support a Protective Role for Lutein and Zeaxanthin in Delaying Chronic Disease . Overview 1 , 2, (2002) 518–524.

[3] J.M. Nolan, E. Loskutova, A.N. Howard, R. Moran, R. Mulcahy, J. Stack, M. Bolger, J. Dennison, K.O. Akuffo, N. Owens, D.I. Thurnham, S. Beatty, Macular Pigment, Visual Function, and Macular Disease among Subjects with Alzheimer's Disease: An Exploratory Study, J. Alzheimer's Dis. 42 (2014) 1191–1202. doi:10.3233/JAD-140507.

[4] L. de Munter, D.H. Maasland, P.A. van den Brandt, B. Kremer, L.J. Schouten, Vitamin and carotenoid intake and risk of head-neck cancer subtypes in the Netherlands Cohort Study, Am. J. Clin. Nutr. 102 (2015) 420–432. doi:10.3945/ajcn.114.106096.

[5] Z. Yilong, Z. Dean, L. Daoliang, Electrochemical and Other Methods for Detection and Determination of Dissolved Nitrite: A Review, Int. J. Electrochem. Sci. 10 (2015) 1144–1168.

531607 A New Surface Functionalized Biosensor for Long-Term *In Vivo* Glucose Monitoring

Abstract:

Diabetes mellitus is a metabolic disorder in which body is unable to control blood glucose level. This disease is a significant and expanding public health problem in USA. According to national diabetes statistics report, diabetes was the seventh leading cause of death in the United in 2015, and 30.3 million people (9.4% of the U.S population) have diabetes until 2017. Diabetes patients need regular and daily blood glucose monitoring to control their blood glucose level. The commonly commercial glucose meters are based on electrochemical sensors. They require fresh blood sample in each test, which can be painful and inconvenient for patients. Considering these drawbacks, developing new fully noninvasive or minimally invasive glucose monitoring devices became the primary need for frequently glucose detection. In this regard, optical techniques including fluorescent spectroscopy, surface enhanced Raman spectroscopy (SERS) are widely used in a number of studies. SERS is a powerful analytical method which can even detect signal molecule. However, this technique is hard to be used for glucose detection since the glucose molecule has low Raman active and affinity with metal surfaces. Herein, we developed a new glucose sensor that addresses the problems in SERS glucose detection. A probe molecule was chosen as an indirect reporter because the SERS signal of the reporter exhibits a glucose concentration dependent behavior on SERS substrate. After the validation of the concept using *in vitro* test, the as-prepared SERS glucose probe was injected in mice skin for *in vivo* continuous glucose monitoring. A commercial glucose meter, OneTouch®, was employed to track the blood glucose variation at the same time. Results of *in vivo* SERS detection shown extremely strong signal and similar trend compared with glucose meter, which proved that this new glucose sensor has great potential for minimally invasive glucose monitoring.

528164 Mobile Technology Based ECL Biosensor Instrumentation

Abstract:

An electrochemiluminescence (ECL) based biosensor was developed utilizing ubiquitous mobile technologies. ECL offers great advantages over fluorescence based measurement: ECL allows high signal-to-noise measurement; ECL is a highly localized and time-triggered detection method; and instrumentation can be minimized as it requires only voltage to trigger the reaction. Cell phone-ubiquity and connectivity creates opportunities to simplify healthcare instrumentation, brings closer to end users, and enable instant data sharing.

A highly compact and affordable instrumentation wad designed and constructed using the mobile technology replacing the traditional ECL detector, computer interface, and data analyzer in an all-in-one mode. Ru $(bpy)_3^{2+}$ was embedded in silicon nanoparticles and sandwich immunoassay was conducted. One antibody is conjugated to the Silicon nanoparticles where Ru $(bpy)_3^{2+}$ is embedded and the other antibody is attached to magnetic bead so that the bound complex can be attracted to the electrode surface separating the false positive signals. Coreactant tri-n-propylamine (TPrA) or 2-(dibutylamino)ethanol (DBAE) was tested for their ability to enhance the signal.

TPrA provided stronger signals for carbon electrodes and optimal concentration was determined for low signal measurements. The challenge of using disposable screen-printed electrode (SPE) was discussed as the binding materials and its geometry in SPE altered the electrode behaviors. The sensing conditions were optimized by experimenting with variables of buffer conditions, trigger voltage, and electrode conditions. The target sensing molecule of vitamin-D binding protein (DBP) was detected at the hundreds of nano-molar level with the optimized conditions.

The sensor demonstrated that a new instrumentation with mobile technology in the point-of-care diagnostics could provide reliability and sensitivity of a high-end equipment. This project is funded by the NSF CBET division.

537782 Electrochemical Detection of Extracellular Bacterial Compounds Using Capillary Electrophoresis

Abstract:

Electrochemical detection of extracellular bacterial compounds using capillary electrophoresis

Martin Kimani, Dr. Edgar Goluch

Department of Chemical Engineering, Northeastern University, Boston, MA

Microbiologists in the research and biotech industry still primarily depend on bacterial culture plates to identify bacterial infections. The bacterial culture plates can take 1 â€“ 5 days before clinicians confirm a bacterial infection and provide the required antibiotic to the patient; this delay in acquiring results can hinder patient care especially with the growing impact of antibiotic resistant bacteria. Therefore, rapid detection of bacterial infections at the point of care would play a key role in improving patient care and potentially prevent bacterial antibiotic resistance. In this work, we present the use of low cost fabricated microfluidic capillary electrophoresis devices coupled with electrochemical detection to detect extracellular bacterial compounds in bodily fluids. These compounds includes electroactive molecules, sugars and proteins which have been shown to readily oxidize at gold, copper and platinum electrodes. We explore if different species of bacterial excreted molecules can be electrochemically distinguished after separation using microfluidic capillary electrophoresis. The portable microfluidic devices are capable of separation and detection under an hour which can greatly improve the time to results whereby clinicians can provide targeted antibiotics instead of awaiting 1-5 days for a culture plate to make the determination of the type of bacterial infection.

The microfluidic devices were fabricated in-house on PDMS and on glass to make 5 centimeter long capillaries at 20 â€“ 75 Âµm diameters. Gold, copper or platinum electrodes were aligned and deposited on glass substrates at the end of the capillary to serve as an end channel detection and for the power supply used for conducting the capillary electrophoresis. The voltage for separation was conducted between 400V to 3000V. The fabricated microfluidic device was optimized to allow amperometric detection of excreted bacterial molecules at the end channel electrodes. Different species of pathogens were studied to determine the selectivity of the fabricated sensors. The advantage of the microfluidic setup is the low cost, portability and ease of use in which a bodily fluid sample can be analyzed with very limited sample preparation. Typical electroactive bodily fluid compositions would be separated out of the capillary by eluting at known intervals therefore allowing for specific determination of additional components in the fluids. In addition, the set-up allows for the analysis of small volumes down to the nanoliter range.

The results from this work can provide the ability to selectively differentiate bacterial species based on the elution profile from a capillary electrophoresis set-up. In addition, this set-up can be coupled with functionalized electrodes at the end of channel set-up to improve sensitivity and selectivity of targeted compounds that may not be readily electroactive on bare electrodes.

520682 Invited Talk: The Next Dimension of Detection: Biomechanical Analysis of Tissue

Abstract:

Since the discovery of the antibody-antigen interactions and the specificity of DNA hybridization, researchers have focused efforts in diagnostic sensor system design around analyzing the chemical properties of biological systems. However, this approach fundamentally limits the type and diversity of data that can be acquired. By expanding this analysis to include the acquisition of other types of data, such as electrical and mechanical analysis, it is possible to greatly increase the data set and sample analysis; thus improving the reliability of the sample analysis.

In the present work, we demonstrate a new approach for analyzing mechanical properties of unprocessed tissue slices. By using polarization maintaining optical fibers as the sensors in a compression testing system, we are able to improve the detection resolution of mechanical analysis of living, recently resected and unprocessed tissue slides. This approach has allowed the mechanical behavior of wide range of living tissues to be characterized, including porcine kidney, pancreas, and liver. The experimental results compare favorably with a visco-elastic model. In complementary work focused on the pancreas, we have designed 3-D printed biomimetic structures based on cross-sectional images of the pancreas. Using compression testing and finite element analysis of these microscale structures, we have identified several characteristic mechanical features, including buckling. When performing similar measurements with the resected tissue using the optical polarimetric system, we observe similar features, indicating that the pancreatic tissue undergoes buckling. This buckling feature can be attributed to the collagen IV network, and is determined to be the dominate contributor to the mechanical response of the tissue. Detection of this behavior in unprocessed living tissue is only possible due to the improved resolution of the optical fiber sensing method. This approach for tissue analysis provides a complementary strategy for understanding biological systems and could enable the development of new approaches for therapeutics.

537557 Rapid Biosensing of Endocrine Disruptors with Cell-Free Protein Synthesis

Abstract:

Human growth, development, and reproductive ability is monitored and controlled by the human endocrine system. Disrupting the balance and control of the endocrine system is linked to reduced IQ, cancer, reproductive challenges, and growth disorders and younger children are especially susceptible. This has led to growing concern over the ubiquitous presence of endocrine disrupting chemicals (EDC) and their effect on humans and animals. A recent study estimated that exposure to EDCs costs the U.S. $340 billion USD each year. This concern has catalyzed biosensor development to detect and study the mechanisms of these chemicals. Although progress has been made, available biosensors remain expensive and time consuming. Here we report a rapid, versatile cell-free protein synthesis platform for biosensing endocrine disrupting chemicals. This technology has the ability to produce a near real-time signal. We also report the robustness of cell free system in combination with a â€œjust-add-sampleâ€ technique to detect endocrine disrupting compounds in environmental and biological samples. Finally, we report engineering the biosensor for improved sensitivity in human samples of blood and urine.

538579 Lab-on-Skin: Epidermal Microfluidic Device for the Capture, Storage, and Colorimetric Sensing of Sweat

Abstract:

The lack of a precision, intimate body interface in modern health related sensing and diagnostic devices frustrates reliable collection of clinical quality data. This presents a tremendous opportunity for developing devices that can fully integrate with the body, for diagnostic, therapeutic or surgical function with important, unique capabilities in clinical healthcare and biomedical research. In this talk, I will present the materials and device designs of skin-interfaced microfluidic system that can capture and analyze biomarkers of creatinine and urea in human sweat for the early detection of chronic kidney disease. Human studies demonstrated the functionality of this microfluidic device in a controlled environment. The results include quantitative values for sweat rate, pH, and concentration of urea and creatinine. The developed skin-interfaced microfluidic device provides new avenues for personal healthcare, medical diagnosis, and biological studies.

522742 Stimulus Response Characterization of an Elastin-like-Polymer Modified Surface for Biosensor Applications

Abstract:

Elastin-Like-Polymers (ELPs) are stimulus responsive polypeptides that change conformation in response to changes in environment such as temperature and ionic strength. At neutral ionic strength and low temperature ELPs are soluble in solution, but, with increasing temperature or ionic strength of solution, ELPs change conformation to form an insoluble protein aggregate. The kinetics of the transition between conformational states is influenced by both the strength of the environmental stimuli as well as the hydrophobic surface area of the ELP determined by the primary amino acid sequence [1]. ELPs are comprised of a pentapeptide repeat sequence [VPGXG] (V = valine, P = proline, G = glycine, and X = guest residue). The guest residue hydrophobicity, total number of pentapeptide repeats and pattern of guest residue selection in repeat sequences determines the hydrophobic surface area of ELP [2].

The stimulus responsive behavior of ELP is commonly investigated for drug delivery applications; however, we propose translating this concept to develop a dynamic sensing platform for biosensor applications [1]. Preliminary investigation includes the immobilization of ELP to a gold surface followed by characterization of two distinct biosensor states. In the extended state, at neutral ionic conditions and low temperature, the ELP are elongated extending from the sensor surface in a brush-like array. The collapsed state occurs in response to changes in environment where the ELP changes conformation and aggregates on the sensor surface. A dynamic ELP sensing platform acts as preliminary development of a biosensor with future works requiring further characterization and incorporation of a biorecognition element for analyte specificity.

The development of a stimulus responsive surface begins with immobilization of ELP on a gold surface using thiol chemistry. Recombinant DNA technology is used to produce ELP allowing for the incorporation of a cysteine residue to provide a thiol functional group. Following surface functionalization of ELP, atomic force microscopy (AFM) can be used to evaluate the extent of ELP surface coverage through characterizing surface topography. Surface cleaning is important to remove any organic matter possibly fouling the surface possibly hindering the kinetics of the thiol gold interaction reducing the extent of surface coverage.

Using surface-immobilized ELP, the extended and collapsed states can be characterized using quartz crystal microbalance (QCM) and electrochemical impedance spectroscopy (EIS). QCM uses a quartz crystal to correlate changes in resonant frequency of a quartz to mass. The shear stress caused by mass accumulation at the sensor surface hinders the ability of the quartz to resonant. EIS quantifies resistance to the flow of electrons, known as impedance, to the sensor surface when an alternating voltage potential is applied. In the collapsed state, a higher impedance is expected with a decrease in quartz resonant frequency compared to the elongated state due to the formation of the protein aggregate monolayer on the sensor surface. The protein layer impedes the electrons interaction with the surface and the resonant frequency of the quartz. Ionic strength, temperature, and hydrophobic surface area will be evaluated as environmental stimuli to trigger the transition of the surface immobilized ELP between conformational states. We will report on our progress on the development of this technology.

The authors would like to acknowledge NSF EAGER (CBET 1638896) for the funding of this work.

References

[1] D. Schmaljohann, "Thermo- and pH-responsive polymers in drug delivery," *Adv. Drug Deliv. Rev.*, vol. 58, no. 15, pp. 1655–1670, 2006.

[2] R. Herrero-Vanrell, A. C. Rincon, M. Alonso, I. T. Molina-Martinez, and J. C. Rodriguez-Cabello, "Self-assembled particles of an elastin-like polymer as vehicles for controlled drug release," *J. Control. Release*, vol. 102, no. 1, pp. 113–112, 2005.

537167 Development of a Unique Dual Ionophore Ion Selective Electrode for the Detection of Proteins and Cells

Abstract:

The need for novel, rapid, and inexpensive biosensor platforms is continuously increasing as the demand for point-of-care capable technology grows. Diagnostic tools which provide results within 30 minutes outside of a central laboratory setting are especially critical for the detection of rapidly progressing, severe medical conditions such as sepsis. Sepsis, which occurs due to a systemic inflammatory immune response to bacterial, viral, or fungal infections, is a major cause of death in the United States. Novel biosensors which allow point-of-care, rapid identification of pathogenic species could dramatically improve patient survival odds by allowing more timely administration of targeted antibiotic therapy. Potentiometric-based biosensors such as ion selective electrodes (ISEs) provide a promising platform for point-of-care detection, as they boast response times on the order of seconds, an electronic signal which can be easily recorded on a simple hand-held electronic device, and inexpensive construction. In this presentation, we will describe the design of a unique ISE modeled after electrochemical potential regulation in living cells and neurons. The dual ionophore ISE (dI-ISE) incorporates two separated ion selective membranes, one sensitive to potassium and the other to sodium. The two membranes are electrically connected by identical inner and outer electrolyte solutions; thus, the equilibrium electric potential developed across the sensor is dependent on the concentration of both ion species near the membrane surfaces, as occurs in living cells. The dI-ISE can be adapted for biosensing applications by the covalent attachment of a biorecognition molecule to the surface of one of the ion selective membranes. Upon analyte binding, the surface of the membrane is partially covered, and resistance to ion transport near the surface of the membrane increases. This causes a shift in the measured potential, such as occurs in living cells or neurons upon closure of selective ion channels. Through the incorporation of the second membrane and ion species, a stable electric potential still exists in the system even when the transport of one ion is completely blocked, allowing a true shift in potential to be measured. Results from studies which simulated protein or cell binding using inert materials such as rubber covers or silicone glue to control membrane surface coverage will be presented. Additionally, a mathematical model based on a modified form of the Goldman-Hodgkin-Katz equation will be presented to describe the surface-coverage based response of the dI-ISE system. Finally, preliminary results from potentiometric and microscopy studies with biotin-functionalized dI-ISE membranes will be presented to show the potential of the system to be used for biosensing applications. These initial studies show significant streptavidin capture on the membrane surface, visualized through fluorescent microscopy, and small shifts in potential after short 30 minute incubations with streptavidin solutions. Preliminary results clearly demonstrate the potential of the dI-ISE for further biosensing applications, including the detection of bacterial and viral species and rare proteins or cells.

534853 Evaluation of Aptamer Technology for Detection of Quorum Sensing Molecules Produced By Pseudomonas Aeruginosa

Abstract:

Aptamers are single-stranded DNA molecules that selectively form a complex with cognate targets by folding into three-dimensional structures. They are becoming increasingly popular as therapeutic agents, but their potential as diagnostic agents remains untapped. Zao et. al., 2013 evaluated eight aptamer sequences, which inhibit biofilm formation of *Pseudomonas aeruginosa* by binding to quorum sensing molecules (QSMs) produced by the bacteria. They identified that ALSAP-5 aptamer had a high affinity (K_D ~20 nM – 35 nM) for a QSM called 3-oxo-dodecanoyl L-homoserine lactone (3O-C12-HSL). Binding of ALSAP-5 to 3O-C12-HSL arrested the growth of *P. aeruginosa* in LB cultures. In this study, we further evaluate the binding affinity of ALSAP-5 for 3O-C12-HSL using three orthogonal biosensing techniques *viz.* Surface Plasmon Resonance (SPR), Quartz Crystalline Microbalance (QCM), and Electrochemistry with the intent to evaluate the performance of aptamer-based diagnostic technologies.

A high affinity binding chemistry of streptavidin-biotin (K_D ~ 10^{14} M) was exploited for attaching the aptamer onto the gold sensor surface. For both SPR and QCM, thiolated biotin (biotin-SH) was bound to the gold sensor surface by gold-thiol bond. Streptavidin was then bound to the biotin surface. Subsequently, ALSAP-5 aptamer was flowed through the microfluidics on to the sensor surface where it bound to streptavidin using the biotin attached at its 5' end. The binding of biomolecules on the gold sensor surface causes change of the refractive index in SPR, which is measured as a change in resonance angle. Similarly, binding of biomolecules on the gold coating of the quartz crystal sensor results in a reduction in the oscillation frequency of quartz in QCM.

For electrochemical sensing, the 5' and 3' ends of ALSAP-5 were functionalized with a thiol group and methylene blue, respectively. ALSAP-5 was immobilized on the surface of the gold working electrode using gold-thiol bond formation between gold surface and thiol group at the 5' end of the aptamer. Electrochemical sensing using square wave voltammetry is based on principle of change in current produced by the redox active molecule methylene blue on the 3' end of the aptamer upon binding of 3O-C12-HSL to the aptamer surface. Binding of 3O-C12-HSL to ALSAP-5 results in a conformational change in aptamer, thereby changing the interactions between the methylene blue and the gold surface, which results in a concentration dependent current change.

Aptamer and 3O-C12-HSL concentrations from 100 nM – 10 µM were tested, and linearity, accuracy and precision, limit of quantitation (LoQ), and limit of detection (LoD) were determined for all three test methods. The results obtained using all three biosensing techniques were comparable.

535299 Chromogenic Ethanol Sensors Enabled By Multi-Stimuli-Responsive Shape Memory Polymers

Abstract:

Ethanol has been worldwide blended with gasoline to make fuels for motor vehicles. In addition, being a good solvent for many organic compounds, ethanol is ubiquitously used as solvent for many commercial pharmaceutical and healthcare products (e.g., antiseptic mouthwash and liquid medicines). Unfortunately, the detrimental side effects of ethanol that greatly impede traditional fuel systems and gasoline engines (especially for aircraft fuel systems) and children health have aroused strong interest in developing inexpensive ethanol sensors for pilots and other people that need ethanol-free products. In this work, a new type of ethanol sensor with good sensitivity and specificity has been demonstrated. This novel sensor integrates the concepts of smart shape memory polymers (SMPs) and two-dimensional photonic crystals. Polymer periodic surface gratings are fabricated by a simple templating approach using optical discs, such as DVDs, as structural templates. This 2D periodic structure can effectively interact with visible light, making the polymer gratings iridescent. By mechanically deforming the ordered gratings, the samples lose their original iridescent colors. The deformed SMP gratings can recover back to their â€œmemorizedâ€ periodic nanostructures when they are exposed to mixtures of good solvents (e.g., ethanol) in bad solvents (such as alkanes). Importantly, different concentrations of ethanol will lead to varied colors, which can be easily perceived by naked eyes and can be used as indicators of the ethanol concentrations in the mixtures. These chromogenic/colorimetric sensors are inexpensive, reusable, durable, and readily deployable with mobile platforms (e.g., a smart phone) for quantitative analysis.

AUTHOR INDEX

Alizadeh, Mahsa 7
Alvarez, M. M. 8
Angelopoulos, Anastasios 42
Arcila, Jennifer A. 16
Armani, Andrea M. 50
Armellino, Donna 36
Badmaarag, Ulzii 42
Bakh, Naveed 45
Balog, Eva Rose M. 53
Berge, Mark 22
Beyenal, Haluk 31
Blenner, Mark 5
Bowman, Christopher N. 37
Buch, Pranali 55
Bundy, Bradley C. 51
Burgess, James 12
Call, Douglas R. 31
Campagnari, Anthony 28
Canty, Mary 28
Cash, Kevin J. 15
Chan, Nathan 45
Chio, Linda 23
Clark, Caelen 34
Clayton, Katherine N. 25
Contreras, Mabel Juliana Noguera 18
Contreras-Naranjo, Jose C. 44
Cruz, Juan C 18
Culver, Heidi R. 37
Dai, Yifan 6
Das, Amitava 27
Del Bonis-O'Donnell, Jackson Travis 1, 23
Doblado, Juan 50
Dong, Qiuchen 47
Dong, Xiaorui 44
Dorfman, Kevin D. 38
Dusane, Devendra 27
Ehrensberger, Mark 28, 29, 34
Elms, Makayla K. 15
Ferris, Mark S. 15
Fraseur, Julia G. 25
Frisbie, C. Daniel 38
Furlani, Edward P. 29
Galvan, Daniel David 3
Garcia-Parraga, Daniel 45
Garcia-Salinas, Pablo 45
Gari, Abdullateef 56
Goluch, Edgar D. 49, 55
Gonzalez, S. L. F. 18
Gonzalez-Gonzalez, Everardo 8
Good, Michael L. 41
Gracia-Rubio, Andres 8
Gustafson, Jenna 39
Haick, Hossam 9
Halpern, Jeffrey M. 46, 53
Huang, Yikun 47
Hudnut, Alexa 50
Hung, Isabella 40
Hunt, J Porter 51
Hunter, Alex K. 7
IV, Patrick Carey 41
Jang, Soohwang 41
Jeong, Sanghwa 1
Jiang, Peng 56
Jones, Kelvin K. 45
Jones, Travis 27
Kepler, Kelly 37
Kiamco, Mia Mae 31
Kilani, Mohamed 17
Kimani, Martin K. 49
Kinzer-Ursem, Tamara L. 25
Kline, Mark A. 23
Kowalski, Wladyslaw 36
Kwon, Hyun J. 48
Landry, Markita 1, 23
Lash-Rosenberg, Lili 50
Lee, Dong Hoon 25
Lee, Hakho 20
Lee, Michael A. 45
Lei, Yu 13, 47
Levicky, Rastislav 40
Li, Xuesong 54
Linnes, Jacqueline 2
Linnes, Jacqueline C. 25
Li-Oakey, Katie Dongmei 11
Liu, Chung-Chiun 6
Liu, Haomin 47
Lochab, Varun 27
Lu, Hang 26
Luke-Marshall, Nicole 28
Luo, Yi 47
Magnino, Sarah 22
Manglass, Lisa 5
Mao, Guangzhao 17
Marco, Vicente 45
Marnoto, Sabrina 46
Marsh, Daniel 48
Martinez, Nicole 5
Martinez-Chapa, Sergio Omar 8
Mastro, Michael A. 41
McGuin, Jana 22
McNerney, Monica 21
McNicholas, Audrey 36
Melvin, Adam 24
Mendoza-Ramos, Jackelin 8
Min, Jouha 20
Moehling, Taylor 2, 25
Mohamed, Abdelrhman 31
Mokhtare, Amir 29
Mollet, Michael 22
Monbouquette, Harold G. 14

AUTHOR INDEX

Morales, Marissa....53
Mrksich, Milan....4
Nguyen, Freddy T. 45
Niepa, Tagbo H. R....32
Onukwugha, Nna-Emeka 32
Osma, Johann F 18
Park, Ji Eun....4
Parry-Nweye, Eloise....32
Patel, Robin 31
Pearton, Stephen J 41
Perilla, A. L. C....18
Peters, Casey 27
Petraitis, Vidmantas 36
Pham, Crystal....45
Phillips, Elizabeth 2
Pinals, Rebecca....1
Pitt, William G. 7
Prakash, Shaurya....27
Prakash, Vijetha Nagendra....44
Raval, Yash....31
Ren, Dacheng....33
Ren, Fan....41
Reynolds, Olivia 54
Rogers, John A. 52
Romo, Luis F. 36
Roy, Sashwati 27
Ruffin, Sade....40
Safa, Nora....24
Sahil, Ahmed Ben....31
Santiago, Grissel Trujillo-De 8
Saraf, Ravi....16
Schaefer, Evan....17
Segura, Crhistian Camilo....18
Sen, Chandan 27
Sikes, Hadley D. 10
Sombolestani, Shayan....24
Song, Donghui 47
Star, Alexander 43
Stoodley, Paul....27
Strano, Michael....45
Styczynski, Mark P. 21
Subramaniam, Vish 27
Swirski, Fillip 20
Tang, Miriam Shakalli....51
Tevatia, Rahul....16
Thakrar, Ami 1
Thomas, Mathew....38
Torgesen, Rebekah N. 7
Treasurer, Eshan 40
Ugaz, Victor M. 44
Vaithiyanathan, Manibarathi 24
Van Duyne, Richard P. 4
Van Wie, Bernard J. 54
Walsh, Thomas J. 36
Wang, Hao 33
Wang, Qiming....50
Wang, Yuan 6
Wang, Yu-Lin 41
Weissleder, Ralph....20
Weng, Mao-Wen....36
Wereley, Steven T. 25
Wilmer, Christopher E. 39
Wintenberg, Molly 5
Wolfinger, Russ 1
Wood, David W. 51
Wood, Ryan L. 7
Wozniak, Daniel 27
Xin, An 50
Yang, Jiancheng 41
Yang, Seung Ook....51
Yao, Jiwei....6
Yonet-Tanyeri, Nihan 4
Yu, Qiuming 3
Yu, Xuecheng....17
Zhang, Yi....52
Zhao, Jing 47
Zmuda, Hannah M. 31
Zuckermann, Ronald N. 23
Zurita-Lopez, Cecilia 50